ENGINEERING EDUCATION AND PRACTICE IN THE UNITED STATES

Engineering in Society

Panel on Engineering Interactions With Society
Committee on the Education and Utilization of the Engineer
Commission on Engineering and Technical Systems
National Research Council

NATIONAL ACADEMY PRESS
Washington, D.C. 1985

NATIONAL ACADEMY PRESS 2101 Constitution Ave., NW Washington, DC 20418

NOTICE: The project that is the subject of this report was approved by the Governing Board of the National Research Council, whose members are drawn from the Councils of the National Academy of Sciences, the National Academy of Engineering, and the Institute of Medicine. The members of the committee responsible for the project were chosen for their special competences and with regard to appropriate balance.

This report has been reviewed by a group other than the authors according to procedures approved by a Report Review Committee consisting of members of the National Academy of Sciences, the National Academy of Engineering, and the Institute of Medicine.

The National Research Council was established by the National Academy of Sciences in 1916 to associate the broad community of science and technology with the Academy's purposes of furthering knowledge and advising the federal government. The Council operates in accordance with general policies determined by the Academy under the authority of its congressional charter of 1863, which established the Academy as a private, nonprofit, self-governing membership corporation. The Council has become the principal operating agency of both the National Academy of Sciences and the National Academy of Engineering in the conduct of their services to the government, the public, and the scientific and engineering communities. It is administered jointly by both Academies and the Institute of Medicine. The National Academy of Engineering and the Institute of Medicine were established in 1964 and 1970, respectively, under the charter of the National Academy of Sciences.

Support for this work has been provided by the National Science Foundation, the Department of the Air Force, the Department of the Army, the Department of Energy, the Department of the Navy, and the National Aeronautics and Space Administration. Additionally, assistance has been provided through grants from the Eastman Kodak Company, Exxon Corporation, the General Electric Company, the IBM Corporation, the Lockheed Corporation, the Monsanto Company, and the Sloan Foundation.

Library of Congress Catalog Card Number 85-61980
International Standard Book Number 0-309-03592-9

Printed in the United States of America

Preface

This report of the Panel on Engineering Interactions With Society was prepared by the panel as input for the deliberations of the Committee on the Education and Utilization of the Engineer. It served as a resource document on the societal, cultural, and historical aspects of engineering for the summary report[1] of the Committee. The panel thanks Mr. Courtland S. Lewis, who acted as rapporteur.

The appendix to this report is "Engineering in an Increasingly Complex Society," which is based on the proceedings of a conference held in July 1983 to examine "issues, challenges, and responses in the history of professional engineering and engineering education." Dr. Arthur L. Donovan acted as conference moderator and rapporteur, and the panel appreciates his efforts in thus helping to provide some of the intellectual foundation for its work.

The panel would also like to thank Dr. Stephen H. Cutcliffe, of Lehigh University, who generously provided a reading list along with a number of key reference works as additional background for the historical sections of the report.

Finally, as chairman of the panel I would like to express my personal appreciation to each of its members for their enthusiastic dedication to the project, which led, I believe, to an interesting and unusual description of the engineering profession and its role in our society.

GEORGE S. ANSELL

CHAIRMAN

1 *Engineering Education and Practice in the United States: Foundations of Our Techno-Economic Future,* National Academy Press, Washington, D.C., 1985.

Panel on Engineering Interactions With Society

GEORGE S. ANSELL, Chairman, Dean of Engineering, Rensselaer Polytechnic Institute (now President, Colorado School of Mines)
THOMAS P. CARROLL, Professor, Science and Technology Studies Division, Rensselaer Polytechnic Institute
SAMUEL FLORMAN, Vice-President, Kreisler, Borg, Florman Construction Company
AARON J. GELLMAN, President, Gellman Research Associates, Inc.
MELVIN KRANZBERG, Callaway Professor of the History of Technology, School of Social Studies, Georgia Institute of Technology
LAWRENCE M. MEAD, JR., Senior Management Consultant, Grumman Aerospace Corporation
M. EUGENE MERCHANT, Principal Scientist, Manufacturing Research, Milacron, Inc. (now at Metcut Research Associates, Inc.)

Consultants

ARTHUR L. DONOVAN, Professor and Director, Center for the Study of Science and Society, Virginia Polytechnic Institute and State University
COURTLAND S. LEWIS, Washington, D.C.

Committee on the Education and Utilization of the Engineer

JERRIER A. HADDAD, Chairman (IBM, Ret.)
GEORGE S. ANSELL, Dean of Engineering, Rensselaer Polytechnic Institute (now President, Colorado School of Mines)
JORDAN J. BARUCH, President, Jordan J. Baruch Associates
ERICH BLOCH, Vice President, IBM Corporation (now Director, National Science Foundation)
DENNIS CHAMOT, Associate Director, Department for Professional Employees, AFL/CIO
EDMUND T. CRANCH, President, Worcester Polytechnic Institute
DANIEL C. DRUCKER, Dean of Engineering, University of Illinois at Urbana (now at University of Florida at Gainesville)
FRED W. GARRY, Vice-President, Corporate Engineering and Manufacturing, General Electric Company
JOHN W. GEILS, Director of AAES/ASEE Faculty Shortage Project (AT&T, Ret.)
AARON J. GELLMAN, President, Gellman Research Associates, Inc.
HELEN GOULDNER, Dean, College of Arts and Sciences, Professor of Sociology, University of Delaware
JOHN D. KEMPER, Professor, Department of Mechanical Engineering, University of California at Davis
EDWARD T. KIRKPATRICK, President, Wentworth Institute of Technology
ERNEST S. KUH, Professor of Electrical Engineering and Computer Science, University of California at Berkeley

W. EDWARD LEAR, Executive Director, American Society for Engineering Education
LAWRENCE M. MEAD, JR., Senior Management Consultant (Senior Vice-President, Ret.), Grumman Aerospace Corporation
M. EUGENE MERCHANT, Principal Scientist, Manufacturing Research, Cincinnati Milacron, Inc. (now at Metcut Research Associates, Inc.)
RICHARD J. REDPATH, Vice-President, Ralston Purina Company
FRANCIS E. REESE, Senior Vice-President, Monsanto (now retired)
ROBERT M. SAUNDERS, Professor, Electrical Engineering, University of California at Irvine (Chairman, Board of Governors, AAES, 1983)
CHARLES E. SCHAFFNER, Executive Vice-President, Syska & Hennessey
JUDITH A. SCHWAN, Assistant Director, Research Labs, Eastman Kodak Company
HAROLD T. SHAPIRO, President, University of Michigan
MORRIS A. STEINBERG, Vice-President, Science, Lockheed Corporation
DONALD G. WEINERT, Executive Director, National Society of Professional Engineers
SHEILA E. WIDNALL, Professor of Aeronautics and Astronautics, Massachusetts Institute of Technology

Staff

WILLIAM H. MICHAEL, JR., Executive Director
VERNON H. MILES, Staff Officer
AMY JANIK, Administrative Assistant
COURTLAND S. LEWIS, Consultant

Government Liaison

LEWIS G. MAYFIELD, Head, Office of Interdisciplinary Research, National Science Foundation

Definitions Adopted by the Committee on the Education and Utilization of the Engineer

Engineer
A person having at least one of the following qualifications:

a. College/university B.S. or advanced degree in an accredited engineering program.
b. Membership in a recognized engineering society at a professional level.
c. Registered or licensed as an engineer by a governmental agency.
d. Current or recent employment in a job classification requiring engineering work at a professional level.

Engineering
Business, government, academic, or individual efforts in which knowledge of mathematical and/or natural[1] sciences is employed in research, development, design, manufacturing, systems engineering, or technical operations with the objective of creating and/or delivering systems, products, processes and/or services of a technical nature and content intended for use.

[1] Including physical sciences.

Contents

Engineering in Society

Executive Summary

INTRODUCTION

The Committee on the Education and Utilization of the Engineer formed the Panel on Engineering Interactions With Society to examine broad questions regarding the functioning of the engineering profession in the context of, and in relation to, American society. Although harder to grasp and quantify than other aspects of engineering education and practice, these topics were considered important because of the enormous extent to which the interests of society and the engineering profession are intertwined. Our economic and social health depends directly on the health of the engineering endeavor, and the health of engineering depends, in turn, on the support of society.

The purpose of the panel's inquiry was thus twofold. First, it examined the impact that engineering and technology development has had on the development of the nation and, correspondingly, the impact of societal demands, values, and perceptions on engineering. The object here was to determine how the engineering community has responded to those societal interests and demands. Second, the panel attempted to assess the structure and development of the engineering profession, past and present, to ascertain whether or not the profession is likely to be adaptable enough to meet current and future national needs.

BACKGROUND

Traditionally, the engineer has been held in considerable esteem in the United States. The concepts of the "heroic engineer" and the "wiz ard" inventor have been a prominent part of American folklore, interwoven with enthusiasm for exploration and development of the land and pride in American ingenuity. But in recent decades the American public has become less enamored of engineers and engineering. A duality of image has developed in which, on the one hand, the engineer is admired for his inventiveness, competence, and practicality; while on the other hand he is often viewed as a corporate "yes-man" of conservative views and little social conscience or consciousness. Mistrust of technology and dissatisfaction with its fruits have become significant new elements in American society. Engineers are seen as having lost their traditional aura of heroism and individuality, to have become anonymous team members, soldiers in the corporate army.

This change in image has important implications for the practice of engineering. Perhaps the new image is exaggerated, but it is nonetheless true that exaggerated images can carry great weight in decision making today, particularly when those decisions are made partly on the basis of public attitudes and opinions. More generally, our trust or mistrust of governing institutions often seems to revolve around these matters. In a very real sense, our society's view of itself continues to be partly tied to its view—whether good or ill—of technology and of our national talent for pursuing it.

For these reasons, the panel focused much of its attention on the historical development of the engineering profession, believing that some understanding of the evolution of American engineering in the societal context is essential for understanding its current structure and status.

Historical Development

Engineering began in America with the building of forts, arsenals, and roads. Engineering for military purposes predominated, but the growing population greatly needed transportation systems, buildings, agricultural implements, public works such as sewer and water supply systems, and machine-made products of all kinds. The first engineers in the United States were European; they brought with them to America their European training and European technology. It was not until after the founding of West Point in 1802 that American-born engineers began to appear. As demand for engineering skills was slow to develop, engineering schools were slow to emerge: For almost the first half of the nineteenth century, only West Point and Rensselaer Polytechnic Institute graduated American engineers.

Civil engineering was the first engineering discipline to attain professional status in the United States. By mid-century, mechanical engi neering had also emerged, as experimentation in machine-shop production of arms, tools, and other implements grew more sophisticated. The central accomplishment of American machine technology in this period was a standardized system for production of parts called the "American System" of manufacturing. This technique, combined with a penchant for innovation and simple, elegant design, began to provide the United States with technological autonomy and to build the foundations of an independent economic strength.

As the population increased and development expanded across the continent, the demand for engineering goods and services continued to grow. To meet these and other educational needs, the federal government began in 1862 (under the auspices of the Morrill Act) to support higher education. This federally subsidized land-grant college system gave great impetus to engineering education, making possible a more scientific approach to technical problems.

As a result, the profession began to diversify. Out of civil and mechanical engineering grew mining and metallurgical engineering. Mechanical engineering became more specialized, and by the beginning of the twentieth century a new emphasis on science in engineering had spawned first electrical, then chemical engineering. Industrial engineering (initially a branch of mechanical engineering) developed to systematize further the manufacturing process—especially in the burgeoning auto industry. Work roles also diversified: While military and independent consulting engineers had predominated earlier, corporations became the predominant force for technology development, and specialized assignments within a project team became the rule. Professional standing, for an engineer, was now very closely aligned with corporate standing.

Wars were strong stimulants to engineering in the United States. Taking World Wars I and II together, government direction of research and development (R&D) for the war effort led to postwar booms in chemical, aeronautical (later aerospace), radio, electronics, nuclear, and computer engineering. Even the Great Depression spurred engineering, through massive government funding of such projects as the Tennessee Valley Authority and the Rural Electrification Administration. Engineering had become the nucleus of the nation's phenomenal productivity and economic strength.

Structural Characteristics

The panel was able to make certain general observations about the internal and external forces that helped to shape the engineering profession in the United States throughout its early development. These early, formative processes gave the profession much of its contemporary structure and set patterns for its societal role, status, and function.

- **Societal Demand for Goods and Services.** On a large-scale this "demand-pull" appears to have been the primary driver of technology development, and particularly of growth in established technologies.
- **Undeveloped Societal Demand.** When demand for a product or a service is latent, entrepreneurs (or, in the present-day context, market analysts) may identify the potential demand and develop the technological means to fulfill it.
- **Technology Transfer.** The availability of new technologies through transfer into a society or from one sector of society to another is another force that sparks demand.
- **Indigenous Advances in Technology.** Autonomous technology development, whether through purposive effort or accidental discovery, can create demand if the new technology answers existing societal needs. This is the "supply-push" factor.
- **Infrastructure Development.** Institutional components must be developed to support the engineering enterprise. These elements are: (a) educational institutions, (b) competitive corporations, (c) research facilities, and (d) technical communication networks.
- **Support by Key Individuals.** It is most often individuals, not institutions, who bring about needed changes in traditional practices and entrenched points of view.
- **Government Support.** Because of the scale of actions needed to foster broad change or development in the engineering profession, government support of and intervention in the technology development process is crucial.
- **Supportive Societal Environment.** There must be a social climate that is conducive to technology development and engineering activity. Key contributory conditions are: (a) societal approval of technological advancement; (b) acceptance by the political and financial "establishment"; and (c) existence of a facilitating market structure.

A key characteristic of the profession has been that it tends to follow quite closely the market for goods and services it provides. Both the individual practitioner and the engineering disciplines are highly responsive to perceived societal demand, although this responsiveness can create problems for engineering education as well as for the engineering employee. Thus, the profession's adaptability is a strong point in that it contributes to economic security, but it is a weak point in that professional engineers are dependent on forces that are largely out of their control.

A related point concerns the great diversification that the response to demand has created among engineering disciplines over time. The existence of numerous separate branches gives rise to a tendency toward narrow specialization in engineers and their institutions (especially in schools and professional societies). Diversity may thus have reduced the cohesiveness of the engineering profession, so that there is less of the sense of shared commitments and values that is found among other well-established professions.

Features of the Present Era

In the period since World War II, the most dominant feature of the environment in which engineering has functioned has been change—rapid, even revolutionary change in nearly every aspect of life and work. In this environment, the impact of all the forces noted earlier has intensified. The panel identified four factors of particular importance for the present-day engineering profession: (1) a great expansion of the roles of government; (2) a rapid increase in the amount of information present in daily life and work; (3) the accelerating rate of technology development; and (4) the internationalization of business and the marketplace.

The large-scale support of national technological, social, and economic objectives by the federal government in the postwar period has led to a variety of new federal agencies. These in turn have led to a boom in the employment of engineers by government, both directly and indirectly, and to the emergence of new engineering disciplines in response to massive government funding of R&D programs. The scale of government-funded programs, particularly in defense, has caused public/defense needs to surpass the private/commercial market as the primary driver of development in engineering.

The major new development in the "information explosion" has of course been the advent of the computer. As a new technology the computer may ultimately surpass the steam engine in its impact on the way business is done and, indeed, on the very nature of business. These machines generate a self-perpetuating demand for the technology they embody. As a result, in the past 15 years there has been a nearly exponential rise in demand for electronics engineers and software and computer engineers, placing considerable stress on the engineering educational system.

The revolution in information products has been both a cause and an effect of the great postwar increase in the rate of technology development in general. The overall rate of technological change has come to exert considerable stress on the engineering system. At the same time, the rise of powerful international competition in nearly every aspect of technology development and marketing increases the pressure. The rate of technology development, the quality of engineering education, and the role of the engineer in society are all far more critical under such competitive circumstances than they were when American dominance of virtually every technical field was secure.

The impacts on the engineering profession are numerous and, in some cases, profound. For example, the trend toward greater specialization has left engineers more vulnerable to "technological obsolescence" in the marketplace. Nevertheless, there has certainly been strong evidence of the profession's adaptability in the face of technological change. The shift from vacuum tubes to transistors to integrated circuits in the electronic engineering field is one instance; the very rapid cross-disciplinary movement into the new aerospace field and, more recently, into composite structures provide two more examples. One reason for this flexibility seems to be that engineering is more interdisciplinary than in the past, so that engineers (while highly specialized) are also able to adopt a "systems approach" to their profession.

The contemporary environment has also placed a great deal of stress on engineering education. The degree of technological change means that schools are unable to keep laboratory and teaching equipment up to date. Fluctuating industry demand brings shifting patterns of enrollment, with great overenrollments in some disciplines. The problem is exacerbated by chronic faculty shortages. Shifts in the economy and in student attitudes also affect enrollment. Schools in general are not well equipped to deal with these fluctuations.

There are also impacts on employment. For example, a growing emphasis on the business aspects of engineering in the postwar period has led many engineers to acquire management training to enhance their professional status and abilities. More generally, the high rate of technological and economic change creates a sense of turbulence in some engineering-oriented industries. Whether there are shortages of engineers or not, this turbulence generates a *sense* of shortage, compounded by the fact that engineers in high-demand fields switch jobs frequently to obtain higher salaries. In addition, with more public attention to technological matters has come an increase in ethical concerns associated with engineering work, particularly in environment-related fields such as the chemical and automotive industries and in the whole area of nuclear energy (for both power generation and defense).

With the expansion of government's role in engineering, significant differences are seen between engineering in government and in indus try. These are primarily due to the basic difference in objectives of the private and public sector organizations: profit making on the one hand, and the performance of public functions and services on the other. The number of government engineers who perform design and development work is relatively small; instead, the majority are primarily involved in the planning and management of contractor services. Most engineers in civil service are also necessarily more attuned to broad social needs and concerns relating to their work than are their counterparts in industry. Finally, there is also a prevailing perception that salaries—particularly in the lower and upper ranges—are lower in government than for comparable positions in industry, and that facilities and support also compare poorly. Because of this image problem, government today has difficulty attracting large numbers of highly qualified engineers.

As was pointed out earlier, the postwar period has also seen a rapid increase in the awareness and public scrutiny of engineering activities by the general public. By the 1970s, changing attitudes had given rise to prevalent "antitechnology" attitudes, deriving perhaps from rising general levels of education as well as the greatly expanded capacity of technology for doing harm to individuals, the environment, and society itself. Engineers have tended to be wary of becoming involved in such politically and emotionally charged questions. However, while antitechnology pressures will ebb and flow, they have become an ever-present fact of life. Engineers and engineering will continue to be scrutinized on the one hand and, on the other, asked to perform miracles.

ENGINEERING AND SOCIETY: THE DYNAMICS OF INTERACTION

Based on its examination of past and present characteristics and tendencies of the engineering profession, the panel attempted to formulate a generalized, informal model of the dynamic interactions of engineering with the larger society. That formulation is briefly summarized here.

Supply and Demand

- The demand-pull factor is the principal driver of technology development and the production of engineers.
- The supply-push of scientific advances is one of the primary stimulants to industry demand for engineers.
- To date, there has been sufficient flexibility in the engineering

supply system to meet societal demand for technology-based goods and services.

- The system has been able to respond to changing demand for three reasons: (1) the engineering educational system is flexible enough to adapt institutionally and pedagogically to new requirements; (2) students react quickly to economic signals in opting to study engineering and in choosing specific fields of engineering study; and (3) change has seldom occurred more rapidly than individual engineers could adapt.
- Engineering institutions reflect the compartmental structure established in the nineteenth century. However, schools have adapted to demands for interdisciplinary engineering study; in addition, intra-and interdisciplinary movement of engineers has not been prevented.
- Use of foreign engineers trained in the United States is another mechanism for meeting demand.
- Because it takes at least four years to educate an engineer, there is necessarily an out-of-phase quality to the time frames in which demand and supply operate.
- In a context of rapid technological advancement and numerous weaknesses in the educational system, it has become increasingly difficult for industry's changing expectations to be met within the confines of the present system.
- Factors that may limit supply response in the future include.

—a demographic decline in the population of 18-year-olds
—variable academic ability of the student pool
—a decline in math/science literacy among secondary-school students
—a drop in the relative attractiveness of engineering jobs in an improving economy.

Maintaining Adaptability

- The focus of the delivery system for engineers is the engineering educational system, where stresses resulting from changes in the nature and intensity of demand are most acutely felt.
- Engineering education is subjected to conflicting pressures for: (1) greater specialization; (2) broader, more general technical education; and (3) the inclusion of more extensive general education content such as liberal arts) in the engineering curriculum.
- The avoidance of technological obsolescence requires that engineers obtain an education featuring a good balance of specialization and breadth of courses.

- Some educational options that afford greater flexibility are:

—emphasis on basic studies in the first two to three years
—five-year degree programs
—cooperative education
—continuing education at home, in school, or on the job.

Managing Change

In terms of its effect on society, automation in the form of computerized systems is the most significant technological change presently in the offing. The issue of technological unemployment may come to have even more negative effects than did the environmental issue.

The outlook is for substantial displacement of workers in both the manufacturing and service sectors, but it is impossible to predict the amount of either. Automation will also create jobs at a substantial rate in both the manufacturing and service sectors, but not sufficiently to offset jobs lost. Computer-aided design and manufacturing systems will likely displace many engineers in the manufacturing sector. Nevertheless, with reduction of the work force in general, engineers are expected to represent a higher percentage of the manufacturing work force than they do now.

Because changes in technology usually bring new industries and new demand, they generally alter employment rather than reduce it. If change is managed well by society, an overall improvement of the quality of life can be achieved. As in the case of environmental problems in the 1970s, the government may have to intervene (directly or indirectly) in labor displacement if the application of technology is to proceed smoothly. What is needed are carefully thought-out social and technological interventions.

OUTLOOK FOR THE FUTURE

In the past, the engineering supply system has demonstrated sufficient flexibility to respond to changing demand. However, changes in the nature and scope of business, in technology, and in societal attitudes and values will affect the demand for engineers and engineering-related products. The elasticity of the supply system will be tested. In addition, unforeseen changes in the engineering environment may further stress the supply system. To acquire some understanding of how the system might function under possible future conditions, the panel proposed a set of hypothetical situations ("scenarios") that would affect engineering to one extent or another. The six scenarios examined were:

1. Continued development toward unmanned factory operation, resulting in the United States regaining world leadership in "smokestack" industries (or, alternatively, losing its competitiveness in manufacturing altogether).
2. Attainment of a recognized capability for commercial utilization of space facilitated by reliable space transportation and permanent in-orbit space manufacturing and laboratory facilities.
3. A major new environmental crisis: large-scale contamination of groundwater resources.
4. Widespread adoption of automated teaching via computer.
5. Rapid shift to use of composite materials as a replacement for metals.
6. Sharp fluctuations in the federal budget for defense R&D.

None of the scenarios examined by the panel appeared to exceed the capacity of the engineering supply system to respond and adapt. But it should be noted that the hypothetical scenarios were examined in isolation, as if each were the only unusual stress being felt at a given time. In reality it is likely that two or more such events would be taking place simultaneously, with combined effects that would be much more difficult to predict and, possibly, to withstand.

Because of the uncertainty about what events—and how many—might occur that would affect engineering, it cannot be simply assumed that the engineering supply system is well equipped to meet any conceivable future. Each of the scenarios would create stress within the engineering community. Even today there are numerous problems of engineering manpower supply, particularly in the area of education. Many of these problems have their basis in societal attitudes toward engineering and technology, or in a lack of public understanding of the technology development process, or in a lack of awareness on the part of engineers of the social ramifications of their work.

Close attention to these problem areas is needed if the interaction between engineering and the American society of which it is a part is to continue to function satisfactorily. Accordingly, the panel directs the reader to the conclusions and recommendations presented at the end of the report.

1

Introduction

It is tempting to view any occupational grouping, whether engineers, lawyers, or teachers—or, for that matter, plumbers or police—as a distinct entity, separate from the society in which it develops and functions. Yet such distinctions, inevitable as they may be, are always artificial. The hard dichotomy thus established is in many ways inadequate for describing the complex, dynamic interactions through which society molds professions and professions shape society. Moreover, the habit of dichotomizing can do damage to the popular conception of a profession and its role within the larger society. This may be especially true in the case of an occupation such as engineering, which is subject to rapid change, much diversity in its makeup, and a considerable degree of mystery (from the standpoint of the general public) regarding the nature of its activities. Under such conditions, it is all too easy for an "us and them" point of view to take root.

With these thoughts in mind, the panel that was formed to examine the broad questions of engineering's functioning within the societal context decided to entitle its report "Engineering in Society." This title is meant to set a prevailing tone appropriate to the symbiosis that exists between the profession and the surrounding culture. It is hoped that, by this means, the discussion will be better able to stress the degree to which the health of the engineering profession and the health of the American economy and society are intertwined.

ENGINEERS AND ENGINEERING IN THE CULTURAL CONTEXT

Traditional Views of Engineering

The popular conceptions of engineering in America have their roots in the founding of the country, in its astonishingly rapid progression from an isolated colonial upstart at the edge of the civilized world to a leading economic power. Those conceptions are interwoven with the tradition of American inventiveness—of "Yankee ingenuity"—and with our popular reverence for such figures as Ben Franklin, Eli Whitney, Thomas Edison, Alexander Graham Bell, Henry Ford, and other practical-minded inventors whose achievements helped to shape the nation. The "can-do" attitude remains an essential part of the American self-image, whether it is applied to landing on the moon or to finding new medical treatments and cures.

Over time, the commonplace view of the engineer has acquired a certain range of definition. On the one hand, he (although the situation is now changing rapidly, the traditional image of the engineer has been distinctly male) is the facilitator of "progress," of economic strength—a builder of bridges, dams, and cities; an expander of transportation, communication, and energy systems. It is largely from this notion that the concept of the "heroic engineer" is derived: the rugged tamer of the wilderness, in his mackinaw and laced boots. On the other hand, the engineer is also the purveyor of technology—of the labor-saving device that shapes home life and the workplace as well as the machine that powers industry. In this incarnation, the engineer feeds America's fascination with the clever gadget, the technically impressive. Here, he is the "wizard," closely allied with the scientist in the popular view.

These laudatory conceptions are by no means universal. In other countries—Great Britain, for example—the engineer is traditionally held in considerably lower esteem, as something more akin to a mechanic or other tradesman (Secretary of State for Industry, 1980). And in the United States, the image of the engineer has proven not to be an immutable one. Changing demographics of engineers may be one reason. Early engineers came from the dominant WASP social sector; but in this century, at least until recently, entering engineers have come to a large extent from immigrant groups struggling to acculturate and achieve status (Noble, 1977). However, a more fundamental reason for the changing view of engineers is that mistrust of technology and dissatisfaction with its fruits—even fear of its consequences—has become a significant new element in American society, one that is kept ever near the forefront of national attention by a vocal minority of Americans.

Thus, in modern times a troubling duality has developed. On the one hand, the engineer is admired for his ingenuity, competence, and practicality. But on the other, he has come to be viewed in many respects as an amoral creature, a corporate "yes-man" of conservative views and little social conscience or consciousness—the calm builder of devastating weapons, the untroubled maker of every kind of environmental contaminant. The panel believes that much of this new duality in the contemporary view of engineers derives from a general confusion of their perceived traditional role with their actual contemporary role in society and the workplace.

The Reality: Diversity in a Complex World

The "heroic" image of the engineer belongs to an era in which the frontiers were physical ones, and daily life often hard; the image itself is specifically that of the civil engineer, in an era in which civil engineering works, whether public or private, predominated. Similarly, the "wizard" concept relates to the early mechanical engineer and (especially) electrical engineer. In both roles, the individual actor was often paramount—or is at least seen today as having been so.

Yet, as we shall see in later sections, these roles are effectively obsolete. The era of the lone surveyor or inventor has long since passed. Engineering has become a collective endeavor, with the engineer most often occupying a place in the organizational hierarchy as a team member. Thus, the traditional view of the engineer's role is complicated by divergent conceptions of military versus civilian engineering, the corporate engineer versus the private consultant, the engineering-school professor versus the industry research engineer, and so on. The picture is further confused by the great variety of disciplines that today comprise the engineering profession. To civil, mechanical, and electrical engineering have been added chemical engineering, industrial engineering, bioengineering, electronics, environmental, systems, petroleum, transportation, aerospace, and nuclear engineering, along with a host of other disciplines and subdisciplines and a variety of analytical and technical fields that are considered a part of engineering.

If the engineer has disappointed, if his halo has dimmed or disappeared, it is because he now lives and works in the same complex and highly stratified world that everyone else in the developed countries inhabits. Most engineers (about 73 percent) today work for corpora tions. Corporate structures, and the practice of modern scientific business management, have relegated many of these engineers to the role of worker—much like the production workers whose role in the workplace they initially envisioned, established, organized, and managed. This is not to say that the engineer does not still perform those functions; in many ways that is the essence of the engineer's role with respect to people, machines, and systems. But the context has changed enormously. There is much more pluralism in the activities of engineers and engineering; the engineer is no longer the individualistic "heroic" figure of American legend. His role (and thus his image) changes as the "product" demanded of him by society changes over time. Whether what is expected of the engineer is invention and development, or efficient production of goods, or improvement of the social milieu, the profession as well as the individual engineer must respond and serve those needs.

Significance of Societal Perceptions

We may well ask whether it is actually important how society views engineers and the practice of engineering. How are engineers and their profession affected by these perceptions, and, conversely, how is society itself affected by its view of engineers and engineering? If there is little effect in either case, then the issue becomes an academic one, of little relevance to a study of the status and future of engineering education and employment, of which this report is a part.

The answer is that these are important issues. Perhaps the simplest way to formulate their importance is to point out that the basic functioning of our society depends on our modern technology; technology in all its forms is by now the indispensable mechanism by which developed nations carry on their economic and social lives. Engineers are, more than any other group, the nurturers and purveyors of this mechanism, this essential product. How society views that product is, in a basic sense, irrelevant; it must and will continue to be delivered. But the perceptions surrounding the product (is it good or evil, necessary or dispensable?) and—by extension—its purveyors, the engineers, can significantly affect the product development process. For example, it can influence the degree and type of support that government gives to engineering education. It affects the numbers and types of students entering engineering studies, and their choice of courses and careers. It alters the direction of research and development by both government and industry, and can result in the curbing of individual lines of technology development through regulation and boycott.

These effects not only have an impact on engineers, they also have strong repercussions throughout our society. The frequent clashing of opposing forces over technological matters is a draining, expensive, and divisive phenomenon. Our trust or mistrust of our governing and corporate institutions often seems to revolve around these matters. To a certain extent, our society's view of itself continues to be partly tied to its view—whether good or ill—of technology and of our special national talent for pursuing it. Therefore, it is important to try to understand how these perceptions evolve and what effect they have. Accordingly, the "image of the engineer" is an underlying theme of this report.

CALCULATING THE VECTOR OF CHANGE: WHERE DO WE GO FROM HERE?

This report will first look back at earlier periods in the engineering story. In so doing it will track the development of various components of the engineering community—not only the disciplines, but the educational institutions and professional societies as well—in terms of the societal interests to which they responded. The object will be to determine how functional the engineering community has been relative to those competing interests and demands: how well the "system" has worked.

The next section of the report will examine the present era, the period since the 1950s, in which many of the previous social, economic, and technological trends and pressures have become intensified. The object here will be to examine the impact of those great changes in scope and scale on the various components of the engineering community, to gain some idea of how well the system is working at the present time.

Based on those assessments of past and present, the next section will construct a generalized, informal model of the dynamic relationship between the engineering profession and the larger society of which it is a part. Finally, the results of this analysis will be applied to an examination of present and potential weak points in the system, focusing especially on a summary of several scenarios that were developed by the panel to project how the engineering system would respond to new stresses.

The report will thus have asked the following questions about the engineering profession and community: Where have we been? Where are we now? Where do we go from here? It seems to the panel that this is a useful—indeed, obvious—way to formulate an inquiry into the way in which engineers and their institutions have functioned and may be expected to function, relative to their social role. It makes it possible to ask whether the engineering institutions are flexible enough, the profession adaptable enough, to function adequately in the modern world.

Much has been made in recent years of the "crisis" in engineering. The term refers variously to shortages of engineering school faculty and laboratory equipment, excessive student populations, inadequate numbers of graduates/ practitioners in certain disciplines, the high rate of obsolescence of technical knowledge and technical professionals, and our declining international competitive posture in certain areas. In any of these cases there is room for argument about whether a "crisis" does in fact exist.

It is partly a question of semantics: What is a crisis? Is it a situation in which irremediable harm will result unless immediate action is taken? If so, what kinds of action? To avoid oversimplifying the issues (and falling into dogmatic traps), this report will address such questions directly whenever they arise—not in terms of "crisis," but in terms of the circumstances and the specific requirements for action. In this connection it may be instructive to read the opening pages of the well-known 1968 report *Goals of Engineering Education* (American Society for Engineering Education), which predicts the technology of "The World of 1984." It is interesting to observe how many of those expectations have not come to pass. One may be led to the conclusion that broad technological change will seldom be as rapid as our imaginations suggest, and, further, that our society and its professional systems may be better able to adapt to change than we might expect. The important thing is, not to maintain a crisis-response posture, but to be aware of the mechanisms and limits of change so that informed choices can be made in a timely fashion.

2

Evolution of American Engineering

This discussion of the historical background of the engineering profession in America represents an attempt to seek out the profession's roots in society. The intention is not to provide history for its own sake, but to determine within the context of historical events and periods whether the engineering "supply" system has been functional or dysfunctional, elastic or rigid in responding to societal demands.[1] Focus will be on development of the major branches of engineering and their supporting educational and professional structures. We will examine selected cases of social interaction and institutional development within these disciplines through the end of World War II and then draw some preliminary conclusions based on that analysis.

DEVELOPMENT OF THE STRUCTURE

Birth of the Technological Society: 1790–1850

The introduction of technology[2] to America roughly coincided with its break away from British political control (Pursell, 1981). This coin

[1] Works listed in the bibliography at the end of the report offer a more extensive and detailed treatment of the history of engineering in America. The appendix to this report provides additional historical information and analysis as well.

[2] "Technology" here refers to the mathematically oriented, machine-based technology that we think of today in connection with that term—as distinct from the handicrafts and making of implements that characterized the technology of Colonial settlers and native Americans.

cidence of two revolutions was caused partly by the rapid growth of technical knowledge and applications taking place in Europe at that time. For several decades after attaining independence, the young nation relied heavily on European engineers and European ideas to conduct its internal improvements projects and to stimulate its fledgling industries. As late as 1816 there were on average only two American engineers in each state (and even these were nearly all self-designated as such) (Noble, 1977).

During the late eighteenth and early nineteenth centuries there were two types of engineering activity conducted in the United States. The most prominent was civil engineering, which encompassed such public works as the building of canals, roads, and forts, and the installation of water supply systems for cities. The second type was what would eventually come to be known as mechanical engineering, but which was at this early stage more accurately described as skilled-mechanic work; typically, a machine-shop owner functioned as producer/entrepreneur for a certain line of metal goods, introducing new techniques as his patrons' needs and his own inventiveness prompted. Of the two types, the civil engineer was significantly more professional in the modern sense, as technical and mathematical training figured more prominently in his background and daily work (Noble, 1977). In addition, the civil engineer during this period had a much broader range of professional involvements. An American engineer such as the British-born and German-educated Benjamin Latrobe, for example, might not only build canals and municipal waterworks, but also design public buildings, dig navigational channels in rivers, and design or direct a variety of industrial establishments (Pursell, 1981). Both types of engineering activity were often prompted by military needs. The drive for continental expansion was inseparable from military aims, and weapons were often a machine shop's largest product line.

Civil engineers also had the first engineering school curriculum offered in America. When Thomas Jefferson established the U.S. Military Academy at West Point in 1802, he encouraged its graduates to devote themselves to public works—to form a corps of civil engineers. For many years this corps was the backbone of American engineering: most railroad engineers, for example, were graduates of West Point (again illustrating the close relation between expansion and the military). However, the increasing scale of civil engineering projects and industrial development throughout the early nineteenth century dictated a need for a larger and more versatile engineering education system (Pursell, 1981). A second school offering the engineering degree did not appear until 1824, when the Rensselaer School (later RPI) was opened; this institute offered manufacturing-oriented training to mechanics and machinists, as well as civil engineering courses. However, there was at the time considerable entrenched opposition on the part of academics to the introduction of experimental science—let alone the "useful arts," or applied science—within the classical curriculum. Consequently, despite an evident need, no additional institutes or technical courses of any real consequence emerged until 1845, when pressure from industry and individual industrialists became strong.

One of the most significant American contributions to technological development came early in this period. Out of the machine-shop culture grew the "American System" of manufacturing based on the production of uniform, interchangeable parts, which was enthusiastically promoted by Eli Whitney and others from 1799 on (Pursell, 1981). As this approach to manufacturing took hold, it made more modern products available at lower cost to more Americans, thus speeding up economic growth and simultaneously enhancing the role of the mechanic/engineer. After the successful completion of the Erie Canal in 1825 there was a rapid increase in economic expansion activities: more canal building, more railroads and machinery industries. Both of these developments increased the demand for engineers and engineering products. The linking of regional railroads (culminating, in the 1850s, in a continental rail network) opened up mass markets and a need for mass production of goods. The Industrial Revolution in America now began in earnest.

As the nation expanded, the mobility of the population increased, especially in a westward direction. The size and number of farms in newly opened areas strained the ability of the thinly distributed population to manage the production of crops. Meanwhile, urban populations were increasing five times faster than the rural population (Pursell, 1981), and the demand for food to be sent to cities over the new transportation networks increased accordingly. These trends led to a severe labor shortage in agriculture—particularly during the harvest, when demand for labor peaked. To meet this need Silas McCormick in 1831 developed the horse-drawn "automated" reaper. Similarly, Samuel Morse pursued a solution to the problem of transmitting messages between cities and across the long distances being opened up by railroads; in 1844 his efforts resulted in the telegraph (the first large-scale and commercially important use of electricity and the forerunner of modern communications).

The development of technology in this early period thus proceeded through the application of available (usually imported) technical knowledge to gradually emerging societal needs. Innovation was a hap hazard process. Development was pushed forward largely through the entrepreneurial efforts of individuals, particularly in the manufacturing area, and societal support for the enterprise of engineering as such was ad hoc and sporadic. It was not until the middle of the nineteenth century that engineering as a profession began to take shape.

Emergence of the Professional Engineer: 1840–1890

The rapid advance of an indigenous technology began by the mid-1800s to produce an identifiable American style, characterized by elegant simplicity of design, efficiency in operation, and ease of production. In 1853, after a London exhibition of many American machine-made products, the British government sent two fact-finding teams to investigate American manufacturing practices (Pursell, 1981). The direction of technology transfer had begun to reverse.

Until this time, science and "technology" had been separate, primarily because of divisions enforced by the colleges, which disdained engineering altogether. By mid-century they had begun to interact. The primary impetus for this change was the growth of larger and more sophisticated manufacturing companies (Noble, 1977). A greater association between science and business led naturally to an increased emphasis on engineering in the industrial context. At the same time, market competition (as well as professional competition for status) was leading to greater specialization among engineers—both the civil and machine-shop variety. The need for a more formalized instructional system than apprenticeship was also becoming apparent. These trends led to increased pressure for schools to provide technical training; at the same time, they began the process of differentiation of engineering activities into formalized disciplines.

The Engineering Education System.

As technical education began to emerge in the late 1840s, it took two forms. On the one hand, established "classical" colleges and universities introduced applied science and engineering studies into their curricula: Union College (1845), Yale (1846), Brown (1847), Harvard (1847), Dartmouth (1851), Michigan (1852), and Cornell (1868). A second development was the evolution of the "institute" schools devoted to technical instruction: MIT (1862), Worcester Polytechnic Institute (1865), and Stevens Institute of Technology (1867) were among the first (Noble, 1977).

At about the same time, government recognition of the importance of technical education to development was increasing. Public pressure for low-cost practical and scientific instruction was also growing, as expressed in popular campaigns such as the "Mechanics' Institute Movement" and the later "People's College Movement" for publicly supported technical universities (Pursell, 1981). These pressures helped to produce the Morrill Act of 1863, which provided for a federally subsidized, nationwide system of agricultural and mechanical (A&M), or "land-grant" colleges. The federal action gave great impetus to technical education. State legislatures and established schools alike eagerly accepted federal grants of land and money, creating schools and departments of engineering. Between 1862 and 1872, the number of engineering schools in the United States rose from 6 to 70. By 1880, there were 85 such schools; and the total of schools and graduates continued to grow steadily for the next 40 years, as engineering partook of a general boom in higher education (Noble, 1977).

Despite these great inroads, engineering retained its "outsider" status in academe. While science (as the experimentally directed outgrowth of "natural philosophy") was gaining slow acceptance as a bona fide element of classical studies, engineering remained more distinctly separate. (It is significant that engineers and other "special school" students were excluded from membership in Phi Beta Kappa by the late 19th century; engineers formed their own honorary society, Tau Beta Pi, in 1885.) Engineering professors experienced this disdain most directly, and it was partly through their desire for greater academic respectability that, after 1870, engineering curricula became progressively more scientific in content (Noble, 1977). At the same time, developments in engineering began to demand the incorporation of scientific knowledge. The focus thus shifted away from the study of mechanical principles, with an emphasis on exercises in shop and field, to mathematical theory and principles of design. To facilitate the increased emphasis on science and mathematics, engineering schools began to build laboratories. This trend was most pronounced in the newly emerging electrical and chemical engineering fields, and had a strong impact on the characteristics of those disciplines as compared to the older branches.

A parallel development arising from concerns about the status of engineers and engineering was the debate over the role of the humanities in engineering curricula. The first institute schools offered nothing but technical courses and were adamant about that fact. Later, schools such as MIT and Cornell initiated concurrent classical studies programs for engineers, and eventually most engineering schools followed suit. In addition, the Morrill Act clearly specified that the "liberal and practical education" of students should include classical studies.

Of course, this admixture was not universally accepted. Many engineering educators (and industry employers) objected to the distraction of students from their technical studies, and to the abstraction and "refinement" imparted by the study of philosophy, religion, and literature—qualities deemed worthless if not dangerous in the future employee (Noble, 1977). However, by the end of the century this view was altering somewhat: the social sciences were gaining general acceptance as additions to the engineering curriculum. This "humanistic-social stem" (economics, political science, sociology, and psychology) was seen as having practical value as more and more engineers became corporate managers. It accommodated a new and broader conception of the professional engineer within an organizational framework.

Diversification of the Engineering Disciplines.

Largely because professional civil engineering education (at West Point and RPI) predated any significant comparable training for other technical occupational groups by many years, civil engineers were the first to acquire formal professional status. By any practical yardstick, civil engineering was a profession in America by the time the great canal and rail projects got under way (around 1820). But perhaps the least ambiguous way to assign dates to the emergence of the disciplines as formalized branches is according to the establishment of professional societies. The American Society of Civil Engineers (ASCE) was formed in 1852. Nearly 20 years later (1871), the mining elements of the profession broke away from the ASCE to form the American Institute of Mining and Metallurgical Engineers, the first of many fragmentations of the profession.

It was not until the last quarter of the century that mechanical engineering emerged as a full profession, gradually evolving away from the role of mechanic in the machine shop. When the American Society of Mechanical Engineers (ASME) was formed in 1880, it was dominated by prominent, established entrepreneurs with powerful business connections. As younger school-trained members—employees of the large companies—entered, what emerged at first was a two-track professionalism featuring a certain amount of tension between these two disparate orientations (Noble, 1977). Gradually, with industrial diversification and greater specialization of mechanical work, the newer, employee aspect of work in this field came to predominate.

In the 1870s, the intensification of business activity and the associated pressure for information dissemination combined with increasing technical advancement to bring about a series of important advances in communications. These included the typewriter (1873), the rotary press (1870s), and the telephone (1876). In addition to the telephone,

Alexander Graham Bell invented in this period the photophone (a system for transmitting sound via light waves), tetrahedral construction techniques, a version of the aileron, and a hydrofoil boat. Similarly, Thomas Alva Edison developed his electric light and power system (featuring the carbon filament lamp) in 1879; by 1885 he had acquired more than 500 patents. George Westinghouse accumulated more than 400 patents during the same period, including his air brake in 1869 (Armytage, 1961). This burst of individual inventiveness, built on the diffusion of the American System of manufacturing throughout industry, brought to a climax the era of the "heroic" engineer/entrepreneur of popular mythology. Devices such as these, and such as the reaper and the telegraph, were very often the product of a single man's inspiration and effort. From the 1880s on, for many engineers invention and development increasingly took on a corporate and collective character. Entrepreneurship continued to be an important force (as it is today), but the proportion of engineers engaged in this type of activity became much smaller.

The first engineering discipline to experience this change was mechanical engineering. As described earlier, there was a lengthy transitional period in which the inventor/entrepreneur/industrialist dominated the profession. Even by the turn of the century, the shop-culture ethos in ASME was still in conflict with the newer science-based, specialization-oriented trends. However, the new engineering environment was given clear expression through the emergence of electrical engineering as a new field. In 1884, engineers employed in the new industries generating and using electrical power broke away from ASME to form the American Institute of Electrical Engineers. This new field had been thoroughly based in science and formal technical training from the start and thus did not have older professional traditions to accommodate. Like the chemical engineering profession that emerged somewhat later (professional society formed in 1908), electrical engineering evolved from science toward technology, rather than the reverse, and was closely identified with the role of the corporate employee. This set the pattern for the future role and professional image of the engineer.

Corporate Technology and the Corporate Engineer: 1880 and After

By 1900, the engineering profession in the United States was second only to teachers in size, with 45,000 members. With the annual output of engineering schools increasing rapidly (up from 100 to 4,500 per year between 1870 and 1916), the growth of the profession substantially outpaced that of the industrial work force and the working population as a whole. Between 1870 and 1916, the relative proportion of engineers in the overall population increased by a factor of 15 (Noble, 1977).

This geometric rise in the engineering work force reflected the great boom in industry as American technology advanced and successive waves of immigrants supplied a labor force and consumer base simultaneously. After 1875, the United States was leading the world in invention and industry. By 1890, it led the world in patents awarded and in the production of iron and steel, coal, and oil. A good index of the acceleration of engineering is the increase in patents given: Between 1790 and 1860, some 36,000 patents were assigned; in the 30 years between 1860 and 1890, there were more than 440,000 (a more than twelvefold increase in less than half the time) (Armytage, 1961). Another index: Between 1850 and 1900, the total consumption of energy in the United States increased fivefold (Pursell, 1981).

In the last two decades of the century, much of this increase in energy, inventiveness, and productivity was harnessed by large corporations. Founded in most cases by inventive entrepreneurs such as Edison, Westinghouse, and Bell, companies like General Electric, Western Union, and AT&T took on a life of their own, absorbing engineering talent and producing engineering products in great numbers for a ready market. Products of the haphazard progress of technology over the previous half-century, such companies now began to make technological progress itself one of their foremost products.

The electrical industry was a major force by 1900, only 20 years after its founding. Just as electrical engineers were setting the pattern for modern professional engineering, their parent industry was establishing new standards for industrial production and management in its development of power generation, lighting, transportation, and communication systems. This industry (1) introduced systematic patent procedures, (2) organized the first industrial in-house research laboratories, and (3) began to provide extensive in-house technical training for engineer employees (Noble, 1977). It was also a participant in the great movement toward product standards from about 1900 on.

Perhaps the most critical innovation was the research lab. At first these emerged ad hoc, in response to some intractable development problem; or they were outgrowths of the company founder's original workshop/lab, such as Edison's Menlo Park establishment in which a team of researchers and technicians worked on development of his electrical lighting system. Later they became indigenous departments of the company, and ongoing R&D became standard for the modern, science-based company. In the process, the research lab (particularly in the electrical and chemical industries) began to blur the distinction between scientists and engineers.

The introduction of in-house training for engineers was also an important new development. With the rapid pace of innovation, by 1900 schools often lagged behind the technical needs of industry—in both course content and school laboratory equipment (still a common problem today). An unofficial cooperative arrangement between academia and industry came into being, in which the prospective employee would receive the more theoretical scientific/ technical education in college and, after graduation, would receive company-specific technical training in "corporation schools," which were a transitional step on the way to professional employment. For the first two decades of the twentieth century this practice remained most common in the electrical industry. In the mechanical manufacturing industries, the experience-trained older engineers continued to mistrust science-based training, and pressured colleges to add "shop training" to their curricula (Noble, 1977). Engineering education in the United States was becoming a major focus of corporate interest and attention.

Another noteworthy innovation of this period was the development of product standards. Pressure for standards began to grow in the early nineteenth century in connection with the American System of manufacturing, as a requirement for mass production. The first standards actually emerged in mid-century (e.g., screw-thread standards were proposed in 1864). But systematic standards did not come into widespread use until the turn of the century, when the American Society for Testing and Materials (ASTM) and the National Bureau of Standards (NBS) became active in this field (in 1898 and 1901, respectively).

Great impetus was given to the standards movement by the railroad industry, which required a standard track gauge along with standard equipment of many kinds, such as safety couplings and air brakes. But recognition of the benefits of standardization quickly spread to every industry, so much so that even standards-setting soon became unstandardized as dozens of corporations, trade associations, and professional societies formed standards for their industries. This situation led the professional societies of the civil, electrical, mechanical, and mining engineers to join with ASTM in 1916 in forming the American Engineering Standards Committee (forerunner of today's American National Standards Institute). Throughout the first third of this century, voluntary standards, developed in large part by engineers, enormously facilitated the manufacture and sale of products, stimulated industries, and spurred the growth of engineering-based companies (Florman, 1981).

By late nineteenth century the growing bond between engineering schools and industry, and the increasing identification of the engineer with his company, posed some problems for the engineering profession. The professional societies were a natural forum for debate on these questions. (Even the ASCE, founded in 1852, had immediately begun to wrestle with "ethics" issues.) Pressure from within and without the professions to standardize the quality of the engineering-education "product" for business needs was one of the principal reasons for the establishment of a Society for the Promotion of Engineering Education in 1894 (Noble, 1977). The central problem was one of conflicting professional identities. Was a professional engineer to be primarily (a) a businessman, (b) an employee, organized along the lines of production workers, or (c) a repository of arcane scientific knowledge?

For many practicing engineers, professional identity centered on the businessman concept. But the interpretation of this role varied among the different branches: In civil, mining, and mechanical engineering it tended to include the consultant and entrepreneurial role; whereas for the electrical and chemical engineering branches (and many mechanical engineers) the focus was on management within the corporate framework. The practicing engineer now found himself in a dilemma analogous to that encountered by early engineering educators, struggling to maintain professional respect and self-respect in an environment not wholly conducive to it. Unlike other professional groups (physicians and lawyers, for example), engineers had become largely coopted by the organizations that their special knowledge, technology, had helped to breed (Layton, 1971). Professional standing, for an engineer, was now very closely aligned with corporate standing. This condition inhered in the nature of the technology development process and was thus inevitable, but it is nevertheless one that continues to be debated even today.

Global Depression, Global War

By 1930, the primary change in engineering was the great scale on which engineering activities were conducted. Industrial research had fueled much of this expansion: From the first industrial research laboratory in 1901 (the General Electric Company's), the number of such labs had grown to 375 in 1917, and to over 600 by 1930 (Pursell, 1981). The rapid growth in the use of electricity and electrical products in the home, combined with the growth and spread of population, created a vast economy dependent on technological goods and services—the "technological society." In addition, new branches of engineering (e.g., chemical and aeronautical) had emerged in strength after World War I.

The most significant new engineering discipline in terms of impact on the economy was production engineering, which was concerned with improving the efficiency of the manufacturing process. An important element was the concept of "scientific management," championed by Frederick W. Taylor and others. These new techniques had their most notable application in the burgeoning automobile industry, where Henry Ford's moving assembly line became the catalyst for revolutionary changes in American life and industry. The effects of the automobile on all the engineering-based industries were profound. The car required tires, radios, engine improvements, synthetic materials, roads, bridges, and fuel. Residential and commercial construction spread far from the city centers. By 1937, U.S. per capita consumption of oil was 10 times that of any other nation (Armytage, 1961).

Across the country, the building of the modern metropolis had enormous implications for engineering. Spearheaded by planners such as Robert Moses, urban development arrived. Skyscrapers, rapid transit systems, and public utilities operating on a vast scale brought a boom in civil engineering in particular. The needs of business for communications and an array of other services were mixed with the requirements of large, densely clustered residential populations. The modern city was becoming a new organism, sustaining a fast-paced, affluent style of living through the provision of a coordinated network of technological goods and services.

Nationwide, the speed of development meant that little was done to coordinate different lines of development, or even to examine their present and future impacts on society and the economy. President Hoover was interested in conservation of resources (land, lumber, and water), and in 1929 commissioned studies that did draw attention to the "unsynchronized" developments in technology. These were clearly matters requiring government attention, but there was as yet little precedent for governmental intervention in economic development on a large-scale. The Panama Canal was one partial exception; and the building of large dams for water management in the Mississippi Valley and the western states early in the century was another step in this direction. Certainly the federal mobilization of scientific and engineering effort during World War I (for example, in the chemical industry) had had an economic impact, if not intent. However, it remained for the Great Depression to provide the opportunity and the rationale for broad, coordinated federal programs bearing on technology.

The Tennessee Valley Authority.

The great experiment in social engineering of the 1930s was the Tennessee Valley Authority (TVA) program. The Tennessee River basin, encompassing an area of some 40,000 square miles, had been subject to recurrent flooding; the river itself, an important link to the Mississippi, was difficult to navigate. In 1933 President Roosevelt established the TVA to solve these and many other problems of the region through a coordinated program based on the construction of a system of hydroelectric dams. Sixteen major dams were built, and five older dams were modified. A 9-foot channel was dredged in the river. TVA provided flood control, power generation, soil conservation, fertilizers, improved public health, and reforestation. This was the largest single construction program ever undertaken in the United States up to that time (Armytage, 1961). It supplied 15 percent of the nation's hydroelectric capacity and 5 percent of the electrical power generated from any source for public use. It reversed the severe erosion in the region, and restored some three million acres to conservation or productive use. Civil and electrical engineers by the hundreds worked on the project, and thousands of other workers were also provided employment. As an example of government mobilization of technological know-how in the service of civilian social and economic needs, the TVA may be unparalleled even up to the present-day.

The Rural Electrification Administration.

An important outgrowth of TVA and the larger government role it portended was the establishment of the Rural Electrification Administration (REA) in 1935. The electrification of the farm had a revolutionary impact on agricultural production, as it provided farmers with low-cost power to light and heat their homes, pump water, milk the cows, and otherwise increase the output that human labor could produce. In addition, it brought urban-style communication to great numbers of Americans and thus broadened the demand for manufactured goods that electrified homes were now equipped to use.

World War II.

Throughout history, technology has had a decisive effect on warfare. World War II was no exception. Even before the United States entered the conflict, it was apparent to the federal government that science and technology should be mobilized to contribute to a prospective war effort. Perhaps the most significant move was the formation of the Office of Scientific Research and Development (OSRD) in 1941, with engineer Vannevar Bush as its director (Pursell, 1981). Research carried out by this agency created the basis for today's "electronic warfare." The war produced such new technologies as radar, controlled nuclear fission, nuclear weapons, the computer, systems theory, jet propulsion, long-range rockets and missiles, synthetic rubber, penicillin, and DDT. It was also a revolution in terms of the scale of technology employed: As a case in point, during the war the Allies used 14 times as much gasoline on an average day as had been used by the Allies during all of World War I. The expansion of research in industry as a result of the war effort was also striking. By 1950 there were 2,700 industry R&D labs in the United States, employing some 175,000 people (Armytage, 1961).

The expansion of industrial research after the war partly reflects the new links forged during the war between scientists and engineers as they contributed jointly to the war effort. One result of those linkages was a greater postwar emphasis on science and mathematics in engineering education. Similarly, the war facilitated the forging of various institutional links among academe, industry, and government, which became permanent after the war ended (the National Science Foundation is one such link, in this case between government and universities. Another key theme of the war was that engineering was recognized as being of critical strategic importance. It was now clear that national security depended on the federal government's maintaining the health of the profession. The work of this panel and its parent committee is evidence of that continuing concern.

The end of the war found the United States in a dominant position globally, with the world's largest and most efficient industrial plant and a strong economy, while those of most other industrialized nations were in ruins. It also found millions of servicemen eager to return home and attend college under the GI Bill. The technological society was about to be inaugurated in earnest.

EARLY STRUCTURAL CHARACTERISTICS OF ENGINEERING

Based on the foregoing examination of the engineering profession as it evolved in America from the late eighteenth century through World War II, the panel made certain general observations about the external and internal forces that helped determine the course of that evolution. The panel recognizes that those early, formative processes may not have direct relevance to present-day events. However, they gave the profession much of its contemporary structure, established inherent strengths and weaknesses, and set patterns for its societal role, status, and function. Thus, a discussion of these factors in the historical context may serve to establish themes useful in evaluating the profession at the present time and projecting its possible future course.

Forces Affecting Development

Societal Demand for Goods and Services.

On a large-scale this "demand-pull" appears to have been the primary driver of technology development, and particularly of growth in established technologies. Demand by towns and cities for municipal water supply systems in the post-Colonial period, for example, was based on the general recognition that such systems were available. Civil engineering expanded through the demand for this and other public improvements; and technology advanced as engineers adapted and improved the associated hydraulic pumps and turbines. Similarly, the need of railroads for a means of message transmission led to the telegraph, which was then adopted as a more general medium of communication. The Civil War intensified the demand for improved transportation and communications systems, leading to a burst of inventiveness that then stimulated business and thus the entire technology development process; electrical and mechanical engineering were specific beneficiaries. High demand for automobiles in the period after World War I is another example of societal demand driving the direction and rate of engineering development. Each particular demand translates into a demand-pull on manpower as well, resulting in the establishment of an educational system or new components suitable for imparting the needed skills and knowledge. But societal demand based on available technology and clearly defined wants should be distinguished from potential, as-yet-unrecognized demand.

Undeveloped Societal Demands.

Often the demand for a product or a service is latent; that is, were a suitable technology available and recognized, demand would appear. In modern times, perceiving these unexpressed needs is often the function of marketing analysts. During earlier periods it was the inventor/entrepreneur himself who identified the latent demand and developed the technological means to fulfill it. Thomas Edison, for example, after an early experience of failure in marketing a device he had invented, always thereafter identified a market before pursuing an idea (Pursell, 1981). The success of the "automated" reaper was likewise due to McCormick's accurate assessment of a need for greater harvesting capacity in the face of a farm labor shortage. Charles Kettering, the legendary director of General Motors' Research Laboratory, owed his phenomenal success to an ability to anticipate the product that "people never knew they wanted until it was made available to them" (see appendix). Once identified and addressed, such hidden needs rapidly translate into demand that further stimulates development.

Technology Transfer.

The availability of new technologies through transfer into a society or from one sector of a society to another is another force that sparks demand. In the early history of the United States, such transfer of technology took place in the form of importation of trained engineers and technical knowledge from Europe—chiefly from England. The flow of technology transfer had largely reversed its direction by the mid-nineteenth century, but remained land remains today) important as a factor in U.S. technology development in certain key areas such as optics, precision instrumentation, and electronics. This factor stimulates demand not only for goods and services, but also for development of an indigenous capability for providing those goods and services.

Indigenous Advances in Technology.

Autonomous technology development, whether through purposive effort or accidental discovery, can create demand if the new technology answers existing societal needs. This "supply-push" factor became especially important in the electrical and chemical industries, where large-scale research was more likely to produce unexpected breakthroughs in science and technology. The panel observes that the potential for such advances to affect the engineering profession is greatest if they are linked to organizational mechanisms by which (a) potential uses of the technology are identified, (b) a potential market can be identified, and (c) demand can be stimulated.

Infrastructure Development.

Extremely important factors in the development of the engineering profession are the components of the institutional infrastructure that supports engineers and engineering. These elements are: (a) educational institutions, (b) competitive corporations, (c) research facilities, and (d) the system of technical communication. As we have seen, engineering education emerged gradually and in the face of resistance from the established academic community. The development of engineering schools was unable to keep pace with technology development and the growing societal need for engineers until pressure from industry and trade groups led eventually to substantial federal intervention and support. Research facilities emerged at the turn of the century as a powerful force for change within the engineering profession. Allied with the expanding influence of science-based industrial companies, they were the greatest stimulant to those engineering disciplines most closely associated with those companies: electrical and chemical. Technical communication, weak and informal in the United States until the Civil War period, did not emerge in any systematic way until engineering schools became established and the professional societies had begun to be active. The dissemination of information on an organized and consistent basis was an essential factor in the burst of inventiveness seen during the 1870s, as well as in the move toward organized research by industry.

Support By Key Individuals.

At a time when there was no coordinated planning or direction of technology development on a societywide basis, the support of influential individuals was critical in the development of engineering as a profession. The efforts of Thomas Jefferson, Stephen Van Rensselaer, Ezra Cornell, and others were instrumental in the initiation of engineering education in America. In the opinion of the panel, it is usually individuals, not institutions, who bring about change in traditional practices and entrenched points of view. When those interested individuals are also in a position to bring governmental and political influence to bear (e.g., Jefferson, Hoover, and Franklin D. Roosevelt), their advocacy is of great importance.

Government Support.

The scale of actions needed to foster increased development in the engineering professions is often too large to be undertaken by individual companies or groups of individuals. Thus, passage of the Morrill Act of 1863 was a pivotal event in the professionalization of engineering, opening up the opportunity for technical training to large numbers of people. It is clear that government support for large public works-style projects such as the Eric Canal, railroads, the Panama Canal, and flood control was crucial in early periods of engineering. Similarly, government action during the Depression and, again, during World War II was in large part responsible for the nation's success in overcoming both of those threats to national well-being by means that were partly technological. At the same time, these actions gave tremendous impetus to engineering in all its forms, providing large-scale engineering employment and fostering the development of high-cost, R&D-intensive new fields such as aerospace and computers. In a technological society, government support of and intervention in the technology development process is crucial.

Supportive Societal Environment.

The existence of a social climate conducive to technology development and engineering activity is also essential. The panel believes that there are three main conditions that contribute to such an environment:

- Societal approval of technological advancement (i.e. is such advancement seen as beneficial?)
- Acceptance by the existing establishment [i.e., do the political,

educational, and economic institutions view engineering activity as a threat to their interests or values?)

- Existence of a market structure that will facilitate the spread of engineering products and the demand for them (i.e., is there a market—whether civilian or government—and a way to reach it?)

Adaptability and Responsiveness

In a market environment, adaptability to changing conditions and responsiveness to social and market needs are healthy characteristics, in general. However, there are certain senses in which these characteristics have negative implications for a profession. It should therefore be useful to examine the extent to which the engineering profession has been adaptable and responsive during its development, and to determine whether these characteristics have functioned as strengths or weaknesses.

One characteristic of the profession, evident in early times as well as today, is that it tends to follow the market for goods and services it provides. It is highly responsive to perceived and expressed societal demand. "Supply-push" is also a significant factor, but this is usually serendipitous and rarely permits engineering to structure and direct demand autonomously. Moreover, once a market is established, a technology is devised, and production is going forward, the system tends to manage output so as to maximize profit. Where there is little new technology development involved, output is often maximized as long as demand continues. (The production of automobiles is a case in point.) This process is stopped only by the drying up of demand, either through saturation or through the obsolescence of the technology. Because demand depends on such factors as competition and economic cycles, it is not always possible to predict accurately what demand will be. Consequently, there is little in the way of an internal "brake" keyed to anticipated changes in demand.

Given these conditions, engineering is forced to follow trends closely—this is true on both a microscopic (the practitioner) and a macroscopic level (engineering disciplines). It means that the educational system has difficulty keeping pace with current trends in demand and technology, and that the "output" (students) therefore always lags external conditions somewhat in skills and orientation. This was a noticeable problem for engineering schools even in the nineteenth century, and today it is part of the basis for a contemporary argument that engineering education should stress basics rather than the trend of the moment.

A strong adaptability to business requirements is a necessary corol lary of the close tie to market conditions. Because by the beginning of the twentieth century most engineers were employees of corporations, the fate of engineers and engineering was strongly identified with the fates of companies and industries. This meant that by that time there was relatively little professional self-determination for individual engineers, and that the professional societies were largely subordinated to the interests and requirements of the industries their members served (see, for example, Layton, 1971).

Thus, the panel finds that adaptability is a strong point in engineering insofar as it contributes to the security and economic survival of the professions. But it is a weak point in that professional engineers are dependent on forces largely out of their control.

Diversity

Much of the discussion thus far has tended to treat engineering as a monolithic, homogeneous enterprise. Yet by the end of World War II, the engineering profession consisted of many distinct disciplines (civil, metallurgical and mining, mechanical, electrical, radio, chemical, aeronautical, automotive, industrial, petroleum, marine, agricultural, and production, or manufacturing, engineering). Each of these branches tended to acquire its own characteristics and its own distinctive orientation toward the practice of engineering, springing from the particular circumstances in which it operated. The existence of separate professional societies for each discipline is one factor. Another is the compartmentalization of engineering schools. The close association of different branches with different industries strongly reinforced this tendency. Thus, the fragmentation of engineering permitted natural differences in personalities, interests, and outlook to become more firmly entrenched.

In the view of the panel, the danger in this great diversity is that it may promote a tendency toward narrow specialization in engineering institutions and among the engineering disciplines. The diversification followed the natural diversification of technologies and product lines, but it meant that a somewhat narrow focus inevitably prevailed throughout an engineer's career. This may have reduced the cohesiveness of the engineering profession, so that there is less of the sense of shared commitments and values that is seen among the clergy, for example, or the military, or the medical and legal professions. However, from a structural point of view diversity is only a problem if it interferes with the profession's adaptability as it develops. One of the purposes of the next chapter is to see whether that has been the case.

3

The Present Era: Managing Change in the Information Age

POSTWAR CHANGES IN SCOPE

After World War II the United States found itself in the role of "leader of the Free World." Its far-flung interests and commitments led it to export funds and technology to encourage development in the ravaged nations of Europe and elsewhere. (The Marshall Plan was the most extensive program of international assistance ever mounted.) The Cold War brought a continuing emphasis on national security, which had ramifications for space and nuclear technology as well as for "conventional" weapons systems—the latter growing more sophisticated each year. At home, the baby boom and a burgeoning economy fueled a massive increase in consumption of goods of every kind, while the continuing expansion of business brought about an accelerating flow of information in the workplace. The concept of change—rapid, even revolutionary change—increasingly dominated domestic and international reality. The time scale of events seemed to become shorter.

In this context of increasing complexity and rapid change, four factors seem to stand out in their importance for the engineering profession: A great expansion of the role of government; a rapid increase in the amount of information present in daily life and work; the accelerating rate of technology development; and the internationalization of business and the marketplace.

Expansion of Government's Role

As we have seen, the federal government had played a key role in technology development in the United States—in continental expansion, in public works and public assistance projects, in agricultural development, and through military systems development. The postwar economic boom was attended by a rapid growth in governmental participation in social and economic processes more generally. A legacy partly of the New Deal and FDR's long reign, federal planning, funding, and direction of major programs was now widely accepted. The large-scale support of national technological-social-economic objectives led to the establishment of new federal agencies: the Atomic Energy Commission in 1947, to pursue peaceful uses of atomic energy; the National Science Foundation in 1950, to support scientific research in many areas of national importance; the National Aeronautics and Space Administration (NASA) in 1958, to develop a civilian space program; the Department of Transportation in 1966, to coordinate expansion and development of the nation's transportation systems.

Perhaps most notable of all, in terms of its impact on engineering, was the establishment of the Department of Defense (DOD) (1949) to coordinate national defense efforts. Military technology development continued at a rapid pace in the postwar period—particularly in the nuclear submarine program, in military aircraft and engine technology, missile guidance and control, and military electronics. Throughout the 1950s and 1960s, the Army Corps of Engineers continued to carry out large-scale development and reclamation projects, particularly focusing on irrigation canals and the dredging of rivers, harbors, and inlets.

Since the late 1960s one aspect of societal demand-pull on engineering has been the development of means of curbing technology itself and controlling its effects. In response to this demand, agencies such as the Nuclear Regulatory Commission, the Department of Energy, and the Environmental Protection Agency emerged to regulate and direct technology development. Large numbers of engineers entered government service or the private sector to work for these agencies directly or under contract to them. The net effect was that engineers now acted as "technological policemen" through the application of engineering skills and knowledge to meet regulatory requirements.

As a result of government funding for R&D in new areas, new engineering disciplines began to emerge, and older ones began to experience a subdivision into new specialties. Massive NASA and DOD spending on aircraft and rocket programs caused a considerable upsurge in the numbers of people engaged in aerospace engineering. Wartime and postwar programs to develop radar, communication, and computer technology, funded especially by DOD, led to the emergence of electronics engineering from the more established radio and electrical engineering fields. Nuclear engineering developed as a hybrid of chemical, electrical, and mechanical engineering to support the late-1950s and early-1960s enthusiasm for nuclear power generation. Transportation engineering grew in proportion with the federal highway system. By the late 1960s environmental engineering was emerging in response to public concern about the disruption of ecosystems and the pollution of air and water by chemical by-products of industry and the internal combustion engine.

These new fields were well funded from the start, and demand for specialists in them would often grow intense over a period of just a year or so. Curriculum development in the new fields as well as the older branches was driven to a great extent by large DOD and NASA contracts for pilot programs and R&D activities, which fed money and requirements back into the universities in the form of research grants. Indeed, in many cases the new disciplines were simply applications of an older set of skills in a specialized setting with enormous funding. It was the degree of specialization and the number of people involved that came to define a field.

Apart from the setting of directions, the major new factor introduced by government support of technology development in the postwar period has been the tremendous scale of programs. The manned space program, defense command and control systems, the interstate highway system, urban development programs, and many other government-funded efforts all represent a quantum increase in the human and technological resources devoted to applying science to societal needs through engineering. The great expansion of the defense industry in particular meant that U.S. leadership in high technology now began to derive from defense rather than civilian needs. This new driver of development in the present era has surpassed the older, strictly commercial market-driven mechanisms for development that characterized the first century and a half of engineering in the United States. Its dominance has become so strong that, in fact, it may be threatening the continued health of those civilian market mechanisms. The panel is concerned that future problems may emerge from either of two directions: (1) a shortage of engineers to meet societal needs apart from those driven by government [e.g., defense and space) and (2) the possibility that government-based requirements will strongly distort the fundamental nature and purposes of engineering education.

To be sure, defense R&D expenditures have stimulated the forma tion and growth of important commercial markets (commercial aviation and computers for business and personal use are just two examples). However, these expenditures have also led indirectly to the decline of interest in fields that later proved important. For example, the near-demise of the traditional electrical power option in engineering curricula had major repercussions when the energy crisis arrived in 1973; and the decline of interest in manufacturing engineering has no doubt figured in the gradual loss of goods production to factories abroad in recent years.

The panel believes that there is a strong imbalance in the overall impact that government spending has on the commercial sector and on defense. Policymakers should recognize that, ultimately, the private/commercial sector and the public/defense sector of the economy are interrelated. To a large extent the nation's economic health, its innovative capacity, and its productivity depend on the strength of private business and industry. In that sense, the strength of the commercial infrastructure is a basic element of national security; its maintenance and support should be matters of concern to the federal government.

The Information Explosion

A second major change in the postwar period has been the emergence of information as a new type of commodity. The technological society produces and uses data at an increasingly rapid rate. The proliferation of technological goods and services combines with the information needs of a growing, increasingly sophisticated population to create a strong demand for improved means of generating, storing, manipulating, and communicating information. Especially in industry and government, problems of information resource management—that is, how to handle and distribute massive amounts of information efficiently within an organization—have gained prominence over the past two decades.

The major new development affecting engineering with regard to this phenomenon has been the advent of the computer. As a new technology the computer may surpass the steam engine in its impact on the way business is done, and indeed on the very nature of business. It is a major factor in the shift toward a service-based economy in the United States, in which the production and management of information predominates over hard goods. Because computer systems, which were devised to handle large quantities of data, also produce it in large quantities, they are both a cause and an effect of the "information explosion" of the past 20 years. Furthermore, advances in computer technology are generalizable to a great many applications, not all of them in business. (An estimated 17 million personal computers were sold worldwide in 1984.] Thus, these machines generate a self-perpetuating demand for the technology they embody. Consequently, there is a great demand for engineers who design and configure computer systems; the 1970s saw a nearly exponential rise in demand for electronics engineers.

A new category of product brought about by computers is software, which instructs the computer in a programmed method of operation. Like any other product, software is designed and developed before being produced for sale. Like many other contemporary products it is highly technical in nature; but it is based on computer rather than physical science (Jensen, 1984). The designing of software products has opened up a new specialty of engineering and is further broadening the definition of engineering work.

Accelerated Technology Development

Fueling the revolution in information products, and to some extent deriving from it, has been a great increase in the rate of technology development in general in the postwar period. Throughout the first half of the twentieth century, technology (whether measured by patents or any other yardstick) had progressed at a steadily accelerating rate. But in the 1950s, spurred by massive government R&D spending, by a vibrant economy, and by mass consumerism on an unprecedented scale, the rate of development climbed to new highs. New technologies spawned new technologies as the demand for engineering-related goods and services continued unabated. The fuller and more rapid incorporation of scientific advances into engineering education and practice quickened the pace of technology development. It became commonplace to observe that the sum total of knowledge was doubling at shorter and shorter intervals.

The overall rate of technological change itself thus had the potential to exert considerable stress on engineering. It is pertinent to ask whether the engineering supply system in general, and the technology development process in particular, has adapted adequately to the high degree of change—and whether it will continue to adapt.

Global Business, Global Markets

Since the 1950s, American business interests have expanded in scope to encompass most of the world's countries. Exports of raw materials, agricultural products, and manufactured goods continue to be a major element of the U.S. economy. The rise of multinational corporations in the petroleum, electronics, machinery, chemical, and other technology-intensive industries, as well as the sale of weapons systems by the government, have a substantial impact on engineering employment and business roles.

The other side of this coin is that many of our allies and many newly developed nations have in recent years acquired (or regained) formidable engineering and industrial production capabilities of their own. Thus, the importation of manufactured goods becomes a major factor for American business and the economy as international competition intensifies. Also, large numbers of American engineers are now employed by foreign multinational corporations and even by foreign countries. Business is effectively becoming internationalized as geographic and language barriers dissolve. The panel believes that the rate of technology development, the quality of engineering education, and the role of the engineer in society are all far more critical under such competitive circumstances than they were at a time when American dominance of nearly every technical field was secure. It is the economic corollary of the earlier assumption by engineering of a critical role in national security. Thus, concerns about American competitiveness, particularly in "high-technology" areas, are bringing about significant changes in the orientation of government toward business. Not only are joint R&D and cooperative industry/university and intercompany ventures being encouraged, but the possibility of targeted government assistance to industries and other forms of intervention is being considered. It is clear that these developments have major present and potential ramifications for engineering.

IMPACTS ON ENGINEERING

The effects of these changes in the scope and scale of American business on the engineering profession are numerous and, in some cases, profound. Because the rate of change is increased and because circumstances often affect more than one industry, impacts tend to cross disciplinary lines and to affect large segments of the profession. If the U.S. economy is no longer isolated from world events, neither are engineers isolated from societywide or worldwide events. One of the purposes of this report is to assess the extent to which the established structure of engineering is taking the strain and meeting contemporary needs. To that end, we will examine impacts on the professional disciplinary structure, on the engineering educational system, on the professional societies, and on the individual engineer.

Multiplying Specialties/Interdisciplinary Activity

The rapid—and sometimes sudden—introduction of new products and processes throughout the present era has caused a fragmentation of disciplines into subdisciplines and narrow specialties. This degree of change (and thus of specialization) leaves engineers more vulnerable to obsolescence. A dramatic example was the substitution of transistors for vacuum tube technology in the mid-1950s, followed by the similar substitution of the integrated circuit for transistors some 10 years later.

Contrary to what might have been expected, the impact on engineers of those two events was relatively minor. In each case, the fact that there were virtually no engineers specifically trained in the new technologies—and that the changes came so quickly—meant that practitioners of the obsolete technology were the best positioned and best prepared to apply the new technology. They adapted.

This capacity for adaptation is often evident when new technologies are introduced. It is even more striking when it involves cross-disciplinary movement. For example, when the manned space program geared up in the late 1950s, there were virtually no qualified aerospace engineers. Instead, aeronautical, mechanical, and electronics engineers, mathematicians, and scientists of all types were able to adapt their knowledge to the requirements of the space-flight regime. When the Apollo program ended rather abruptly in the early 1970s, those several thousand engineers were eventually reabsorbed by industry—although the process was traumatic for at least three years, and its repercussions may still be seen in the careers of individual engineers.

Currently, new composite materials being employed in the construction of aircraft bodies require "composite structures engineers"; since there are few people actually trained in this technology, the need is being met by metallurgical engineers, materials scientists, and chemical and mechanical engineers.

One reason for this capacity for flexibility may be that engineering work is often more interdisciplinary than in the past and is becoming even more so. This might seem paradoxical, given the increased specialization mentioned earlier; but in reality, specialization often demands the presence of many specialists in different fields on a development project, particularly for complex systems. Thus, engineers acquire on the job a familiarity with associated or related specialties, as well as added competence to handle real-world problems that are beyond the scope of any narrow group of skills.

These countervailing requirements to be a specialist and a generalist are part of what is, in effect, a new definition of engineering. The new definition derives from a pervasive trend toward the systems approach to engineering development. The aerospace field led the way in developing the systems engineering approach, because of the emphasis on high performance at minimum size and weight. In general, systems engineering permits the interfacing of various subsystems and components of a complex product in such a way that performance, weight, cost, and other important parameters can be optimized in selective fashion. The product can be designed as a single, integrated system, rather than as a loose assemblage of separate systems.

The interfacing of different areas of knowledge is also essential in new fields such as biotechnology, in which sophisticated scientific methods are used by engineers for production of completely new forms of biological "materials." Even as conventional a project as the design and construction of a modern office building is an exercise in the systems approach; heating and air conditioning engineers, structural engineers, design engineers, electrical, electronics, and environmental engineers routinely participate with civil engineers and architects in the development of a building that functions in many respects like an animate object. The panel believes that such a working environment imparts a flexibility to engineers that allows them to better adapt to the changing environment in which they operate.

The Educational System

The rapid pace of technological change, the increased degree of specialization, and sharp fluctuations in demand for engineers in various fields have all placed considerable stress on the engineering education system. Over the past 10 to 12 years, as the overall number of students entering college has plateaued and federal subsidies have begun to decrease, engineering schools have had fewer funds available for improvements to existing facilities and equipment—even though at the same time engineering school enrollments have climbed dramatically. Rapid changes in industrial equipment and tools used by engineers—particularly in electronics engineering, but also for computers in general—have meant that schools cannot afford to keep current the equipment they use for training engineers (see, for example, National Academy of Engineering, 1981). Thus, in the most rapidly developing and critical fields, graduates enter industry with a serious lack of some important skills and knowledge.

High salaries and attractive benefits offered by industry to young B.S. engineering graduates have led to a severe decline in the number of American students opting for graduate study in engineering—especially at the Ph.D. level. Consequently, there is a shortage of Ph.D.

engineers to staff engineering schools. As a result, schools have difficulty coping with larger enrollments and shifting patterns of enrollment. With employment in industry booming, a relatively low-paying faculty position is less attractive to qualified young engineers. More money is not the only consideration here; the nature of the job in general is less appealing under today's constrained circumstances. The shortage of faculty has been a major problem for engineering schools for a number of years (see, for example, Shakertown Conference, 1981). Combined with the generally increased numbers of engineering students in classes, the changing patterns of enrollment, and the scarcity of adequate equipment, the faculty shortage has serious implications for the quality of engineering graduates (see National Association of State Universities and Land Grant Colleges, 1982).

Fluctuating demand by industry for graduates in various fields and with specific kinds of training is something that schools in general are not well equipped to deal with—particularly when changes in demand occur relatively quickly. Since the duration of schooling is generally four years, there is a lag time of at least that long before requirements can begin to be met. The high demand for environmental engineers came somewhat suddenly around 1970; some seven or eight years later, that demand declined just as abruptly. Fortunately for many young environmental engineers who had just entered the profession or were still graduating at that point, their training was sufficiently interdisciplinary (usually chemical and industrial engineering with some chemistry and biology on a civil engineering base) that they were still employable by government and industry in other areas (for example, energy systems, safety, occupational health) if environmental jobs were not available. However, not all environmental engineers were generalists and thus so adaptable. And in other disciplines, where greater specificity of knowledge is the rule, such flexibility is not as easy to achieve.

In fields where growth is forestalled by stabilized or declining demand, surpluses of engineers occur. At present, for example, civil and chemical engineers are said to be in oversupply. This condition is partly a function of increased demand in other fields—intensive development elsewhere draws capital resources as well as consumer interest away from mature industries. Here again, these shifts often occur more quickly than the student cohort is able to adjust to them.

The example of environmental engineering suggests another form of fluctuating demand that has come to affect engineering education in the past 20 years: fluctuations in student demand for engineering as a major. The late 1960s and early 1970s saw a dramatic drop in engineer ing school enrollments, resulting from a decline in general economic activity, a recession in the aerospace field, and changing attitudes among the young. Yet student demand for engineering education later rose as sharply as it had dropped: Fluctuations at this end of the "engineer supply system" can create stresses as great as fluctuating industry demand can create. Figure 1 depicts changes in engineering enrollment, and their primary causes, over a nearly 40-year period.

Engineering schools and departments of engineering have to cope in different ways with both of these stresses, usually under conditions of declining resources and diminishing faculty. This is not an easy task; it has led to calls of "crisis" from many quarters in recent years. Fortunately, government and industry are now paying attention to the seriousness of these problems and to the need to devise ways of easing the strain on the educational system. Industry, for example, as an alternative to hiring engineering faculty members, has begun to emphasize such creative approaches as shared staffing, fellowships to encourage graduate study, support for young faculty, and "forgivable" loans. Cooperative industry/university R&D programs in such fields as manufacturing engineering, robotics, and computer-aided design and manufacturing are also a positive step.

The Professional Societies

Much of the pressure to manage change in the present era has been put on the engineering professional societies. The role of the societies has largely shifted, over the last 50 years, from that of a business information clearinghouse (in essence, a club) to that of an educational society. The societies are all active in publishing technical papers, sponsoring conferences, etc.; through technical communication they follow advancements in the state of the art. To some extent they also function as spokesmen for the interests of their members in the policy-making process (whether state or federal).

A third, and very important, function is their participation in the voluntary standards-setting process for techniques and products relevant to their respective disciplines. Relying on member support and participation, societies develop standards and submit them to the American National Standards Institute (ANSI) for authentication and publication.

A fourth function of increasing importance for the societies is representing the engineer to the public at large. This public relations function is relatively new, deriving from the late 1960s and early 1970s, when mistrust of technology was more prevalent in society. In essence,

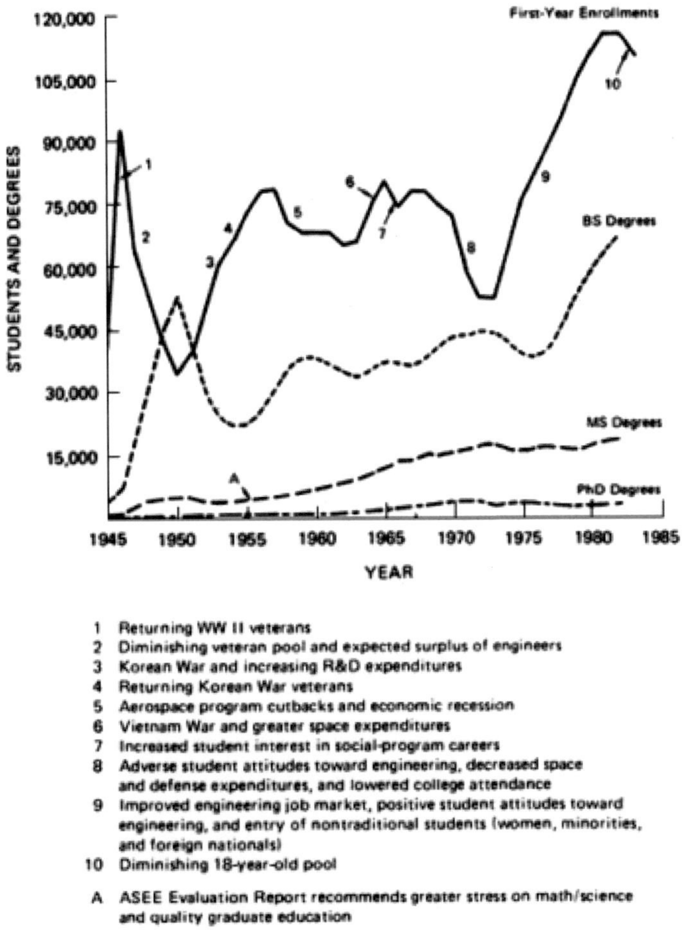

Figure 1
Engineering degrees and 1st-year enrollments: Historical factors influencing changes in engineering enrollments.

it is an attempt to represent the profession accurately to the voters and taxpayers whose support for engineering and for technological advancement in general is important to the profession. A fifth function of the societies is related to this concern for image, although it predates it considerably: The professional societies are active in the continuing process of establishing and adjusting professional ethics. The historical basis for this concern is the duality of the engineer's role as both professional and employee (Florman, 1981). The issue has intensified in the present era as the potential harmfulness of many engineering products has increased (particularly in the chemical and nuclear engineering fields), and as public attention to these matters has grown accordingly.

The Engineer as Employee

Engineer as Corporate Employee.

In the postwar period the rapid growth of big business has led to major changes in the way that most engineers work. A growing emphasis on the science of business administration from the late 1950s on has strongly affected the role of engineers in the corporate world; indeed, many top engineers nowadays acquire management training to enhance their professional status and abilities. Panel members now see indications that, with increased international competition in recent years, the emphasis in management style within many companies is shifting toward the integration of technical knowledge with management skills.

The more competitive and international environment of engineering today has multiple impacts on the engineer as a corporate employee. A variety of new business management approaches have come into use in engineering-oriented companies during the last 10–15 years. One of these is the "matrix management" structure for organizing project work. Under this system, engineers, scientists, and technicians are assigned as needed from functional departments for the duration of a project; when the project is concluded, the project team is broken up and dispersed to other projects. While this approach permits efficient allocation of human resources, in many cases it minimizes the cohesiveness of the team because members do not work together on a permanent basis (of course the length of association depends on the size of the project). Such project teams also usually include a large number of engineers, so that specialization of individual roles is emphasized. This may again detract from an individual's sense of professionalism and commitment to the project.

Rapid developments in technology and the changing competitive fortunes of companies create a sense of turbulence in some engineering fields—particularly in the high-tech electronics, aerospace, and biotechnology industries. Whether there are shortages of engineers in these fields or not, the sense of shortage persists. The problem is compounded by engineers in these disciplines frequently switching jobs to obtain higher salaries. This practice imparts a "free lance" quality to contemporary engineering employment in many fields: the emphasis is strongly on the engineer's personal advantage and advancement, often at the expense of company welfare. The loss of company identification that results from this mobility complements the loss of team identification that may result from project staffing practices.

Another important aspect of engineering work life in the contemporary corporate environment is the tension that many engineers feel between their professional role and their role as an employee. This tension has been present to some extent since the late nineteenth century, when corporate employment of engineers became widespread; but it has acquired new forms with the intensification of business competition and the development of potentially harmful commercial and consumer products. The most common form is the emergence of ethical dilemmas such as the question of "whistle-blowing." These situations often involve instances of blatant wrongdoing, where one's duty as a citizen as well as a professional is clear-cut. But there are also more subtle ethical questions that a professional must sometimes confront, relating perhaps to a basic conflict between one's values and the nature of one's work on a particular project.

Engineer as Government Employee.

The engineer as civil servant is not a new phenomenon, or even a phenomenon strictly of this century. One of the earliest examples of the engineer as employee on a large-scale was the Army Corps of Engineers, and planners of development on the municipal, state, regional, and national level have often been engineers. However, it was not until the 1930s, and particularly from World War II on, that government began to employ civilian engineers in large numbers from every discipline. In the postwar period the formation of the various federal agencies dedicated to planning, directing, and regulating development in nearly every area of social and economic life prompted a virtual boom in engineering employment opportunities. By 1980, government employees at every level of government accounted for 15 percent of the 1.4 million engineers then in the U.S. work force (unpublished NSF data). Table 1 shows the distribution of these engineers in the federal government, the military, and state and local governments.

Apart from direct employment, government supports many more

engineers indirectly, through contract funding. At the level of prime contractor, the federal government supports an additional 24 percent of all U.S. engineers; subcontracting adds another 8 percent to the total (based on estimates provided by Dr. Aaron Gellman).

TABLE 1 Engineers in Government, 1980

Category	Number Employed	% of Total
Federal	101,600	7.3
Military	22,300	1.6
State & local	84,300	6.1
All government	208,200	15.0
Total U.S.	1,387,000	100.0

Source: NSF, unpublished data.

Engineering in government is different in a number of significant ways from private-sector engineering employment. The primary difference has to do with the nature of the employer. Because government is noncommercial and nonprofit, many of the features of work life that predominate in competitive industry are absent, or at least not as prominent, in government engineering employment. The number of government engineers who perform design and development work is relatively small, according to estimates given to the panel by personnel officers of various mission agencies. Usually these "engineering" engineers are associated with testing and standards-setting activities—except in the military, where a considerable amount of systems development is done by (usually civilian) engineers in the different services.

Instead, the majority of engineers across all categories of government are involved to a great extent in the planning and management of contractor services. Thus, the managing of budgets and schedules and the competition for fiscal resources form a considerable and distinctive part of engineering work in government. This contrast between engineering in government and in industry stems from a basic difference in the objectives of the private and public sector organizations: profit-making on the one hand, and the performance of public functions and services on the other.

An oft-cited aspect of engineering in government is the perception that salaries are lower than for comparable positions in industry. Research and development facilities are also often believed to be less advanced and less complete than in industry; office space and support services are another area in which government engineering work is often considered to compare poorly with engineering in the private sector. Whether true or not, these perceptions contribute to a prevailing belief among engineers (and other professionals as well) that government employment is comparatively unattractive. Because of this image problem, government today has difficulty attracting large numbers of highly qualified engineers. And because of the very real inducements of industry employment, it also has trouble keeping experienced personnel. By and large, there is a unidirectional flow of engineers out of government and into industry—particularly in the federal government/military, and most particularly for those whose work has involved them in state-of-the-art development projects in electronics, computers, and other growing fields.

This loss of experience and talent from the government work force is, in one sense, unfortunate; but it may also be beneficial in that certain positive values gained in the service of government are thereby continually being circulated into industry. These values derive from the third way in which engineering in government differs from engineering in the private sector; that is, most engineers in civil service are necessarily more attuned to broad social needs and concerns relating to their work than are their counterparts in industry. In many federal agencies they stand to some extent as intermediaries between economic forces and the greater public good, through regulation of industries, setting of safety and quality standards for industrial products and practices, and enforcement of those standards through testing. At the state and local level they also represent the more specific interests and needs of the people in the jurisdictions they serve for the entire range of government services. As the role of government has expanded, as regulation of private-sector activities has increased, and as general public interest in issues such as the environment, nuclear power, product safety, and government spending has intensified, this aspect of the government engineer's work has become proportionately more demanding.

Intensification of Social Issues in Engineering

As we have seen, an indirect effect of the changes in scope and scale of engineering activities in the postwar period has been an increase in the awareness and critical scrutiny of these activities by the general public. By the 1970s, changing societal attitudes had given rise to a prevalent mistrust of technology—often referred to as "antitechnology" sentiment (Florman, 1981). This change from the sanguine attitudes of earlier periods has been partly the result of rising educational levels in the population as a whole since World War II, so that there is less awe of the engineer, less willingness to trust engineering implicitly and to accept on faith the value of engineering achievements. After all, the engineer is just another college graduate. Heightened critical awareness is also a function of the greatly expanded capacity of technology for doing harm to individuals, the environment, and society itself. While popular attitudes toward technology in general have become considerably more positive in recent years (Yankelovich, 1984), criticism of particular projects and programs is still often in evidence.

Although antitechnology sentiment could be detected in the early part of this century (as in Chaplin's film "Modern Times"), the growth of social concerns regarding engineering activities in the present era can probably be traced from the atomic explosions that ended the war with Japan. Those events, effective as they may have been in ending the war quickly, were an appalling revelation of the power of science and engineering working in tandem. The environmental effects of industrial and auto emissions into air and water became a major issue during the late 1950s and early 1960s, made evident by urban smog and dying rivers, and publicized by books such as Rachel Carson's *Silent Spring*. Underlying public concerns about technology and the morality of its purveyors increased during the Vietnam War, with its televised scenes of napalmed villages and defoliated jungles. During the same period, Ralph Nader projected questions about the responsibility of manufacturers in the design and production of consumer goods into the public consciousness. Later in the 1970s, Three Mile Island brought latent fears about the safety of nuclear power to the fore, further curbing development of that already struggling industry. Currently, the effect of automation on employment in large manufacturing industries is becoming a major social issue.[1]

The other side of the antitechnology coin is that with greater public awareness of the power of technology to shape society has come a new set of demands for technology to improve life. There are constantly rising expectations for better performance, reliability, and safety of products. We demand economic growth but expect technology to maintain a clean environment. We look to technology for the means to minimize the danger of war: inspection techniques, warning systems, etc. We want engineers to make us invulnerable—that is, to ensure that we can win any war—and at the same time we require that they provide

[1] A lawsuit in the California courts as of the time of writing is a case in point. The suit challenges the right of California state universities to pursue research in automation, on the grounds that public funds are being used to further corporate interests to the detriment of workers—the "public." The suit charges that such activity is in basic conflict with the intent of the Morrill Act.

the technical means to prevent war. We expect medical benefits from biotechnology and new or extended energy sources from chemical and petroleum engineers.

And, in fact, engineers and the engineering-related industries meet nearly all of these expectations. It is undeniable that without the technological advances made and implemented just since World War II, Americans would not be as well off as they are today. Without all the technology that supports our large population and modern service-oriented economy, the standard of living and the quality of life in the United States would both be lower. People would generally have less mobility, less leisure time, less entertainment, less time for education, less enjoyment, a less reliable food supply, a dirtier environment, and shorter lives. Yet with many technological advances comes a backlash. Effective detergents containing phosphates turn out to produce "bloom" on ponds. Cleaned up and lengthened industrial smokestacks turn out to cause acid rain. Engineering is required to solve these problems, too (and, ironically, is held partly to blame for them).

What are the implications of these social concerns for the practicing engineer today? Antitechnology tides have ebbed and flowed throughout the twentieth century, but it is likely that engineering and technology will continue to be scrutinized and criticized on the one hand, and, on the other, asked to perform miracles. Engineers will have to learn, at least to some extent, how to operate in a fishbowl. Government engineers have for some time been aware of how intense this pressure can be. The panel suggests, then, that one new requirement may be for engineering education to prepare engineers to conduct their professional activities with a greater awareness of their social responsibilities. They should be trained to view their work in light of anticipated criticism—not just from a technical standpoint, but on a social basis as well.

There are obvious problems inherent in this—beginning with the fact that, in industry, individual engineers have rarely had control over whether or not a given line of development is to be pursued. Once a decision has been made, usually the engineer's choices are regrettably well defined: participate or leave. But if more engineers move into corporate management, their influence in such matters will grow. In addition, if the majority of young engineers become sensitized to the social ramifications of their work during the course of their education, their collective viewpoints may come to represent a formidable force within their respective industries. This would indeed be a powerful demonstration of the exercise of professionalism and professional responsibility in the modern engineering context.

The engineering profession as a whole has tended to be wary of becoming involved in broad social questions relating to engineering work (see Christiansen, 1984). For one thing, such issues are often highly charged politically and emotionally, and full of ambiguity. As such, they are not very compatible with the rational, pragmatic style of mind that characterizes the engineer. For another thing, such issues tend by their nature to threaten the stability and security of the corporate and commercial world in which most engineers work. But concerns of this kind are increasingly impinging on the professional ethics of engineering. And, as was just pointed out, they may do so increasingly in the future.

The panel believes that it is entirely appropriate for engineers and the engineering profession to formulate reasonable views on these matters—in fact, professional responsibility requires it. Armed with the pertinent facts and a broad view of the world around them, engineers should find that they can apply the engineering problem-solving approach effectively even to nonengineering problems. Certainly the professional societies, which have long grappled with ethical questions, can be instrumental in informing engineers and addressing large political and social issues on behalf of the profession. One logical mechanism for accomplishing this could be an umbrella organization like the American Association of Engineering Societies (AAES), working in concert with the various professional and technical societies. Whatever the best means to meet it, the need for the profession to acknowledge and respond to social issues will continue to grow stronger.

4

Engineering and Social Dynamics

In previous chapters we have examined the development of the engineering profession in America and drawn some tentative observations about the nature of its actions and reactions, in earlier periods as well as recent times, with respect to the larger society of which it is a part. In this chapter we attempt to consolidate those historical characteristics and tendencies into a more generalized model of the dynamic interactions of engineering with the larger society. We discuss the effects of those interactions on the profession and society as a whole, and attempt to establish some key areas where functional problems may exist now or in the future.

FLUCTUATING SUPPLY AND DEMAND

The Societal Demand-Pull Factor

A principal driver of technology development is societal demand for goods and services. Furthermore, an advancing technology itself tends to stimulate demand, if the technology accords with existing societal needs. Societal attitudes toward engineering and technology development also have a major impact on the type and level of demand for engineering-related goods and services. The demand for technological goods and services translates into demand by industry and government for engineers in different disciplines. This is the "demand-pull" factor.

Industry is highly specific about the kinds and mixes of skills it requires in engineers it wishes to employ. Yet the nature of these demands changes rapidly in response to the changing business, technological, and general economic environment. Substantial changes in the pattern of government demand —particularly in the defense area—are increasingly a major factor. In a context of rapid technological advancement and numerous weaknesses in the educational system, it has become more difficult for industry's changing expectations to be met within the confines of the present system. Therefore, there are movements in the direction of industry's modifying its demands or joining with schools in an effort to improve the quality of the supply of young engineers.

The demand-pull for engineers and engineering products is quite different from the "supply-push," which is the principal driver for scientists and scientific research findings. Indeed, the supply-push of scientific advances is one of the primary stimulants to industry demand for engineers. This difference in motivations and dependencies is a major factor in the different societal perceptions (and professional roles) of engineers and scientists.

Mechanisms for Meeting the Demand

There are serious questions about whether the educational system, organized along disciplinary lines that were formed in the nineteenth century, is adequate for responding to today's business and technical problems. The same nineteenth-century divisions are reflected in the professional societies and associations, reinforcing the compartmental nature of engineering.

The compartmentalization found in engineering institutions suggests that it would be difficult for new disciplines to develop in response to new societal demand. But this has not been the case. Hybrid fields such as environmental, nuclear, aerospace, and computer engineering have emerged rather quickly to meet demands in recent decades. There was little resistance by the established educational infrastructure. In practice, engineering schools were eager to accommodate the new growth areas. Among practicing engineers there has been considerable movement across professional boundaries to meet the needs of an emerging technology—as seen in the aerospace field and, most recently, in the composite structures area.

Apart from internal adjustments, another mechanism by which the supply of engineers is adjusted to meet demand is the use of foreign engineers, trained in the United States, to fill shortages. This is particu larly true in the case of Ph.D. engineers, since a disproportionate number of current U.S. doctoral candidates are foreign nationals.

There is a fine line between shortage and surplus of engineers. To a great extent the existence of either one is a matter of individual perception. But any deviation (real or perceived) from a balance between the two tends to cause turbulence in the profession and in industry. This problem is intensified by the fact that demand tends to alter more quickly than supply can be adjusted—it takes at least four years to educate an engineer. Thus there is necessarily an out-of-phase quality to the time frames in which demand and supply operate.

By and large, however, there has been sufficient flexibility in engineering education, and in the profession as a whole, to meet past needs. Yet there have been significant changes in societal attitudes and values, as well as in the nature and scope of business, that will affect the demand for engineers and engineering-related products. The elasticity of the supply system will be tested. It remains to be seen whether it can continue to function adequately under current and future conditions.

Factors Limiting Supply Response

In an assessment of the adequacy of the engineer supply system a number of important variables come into play. One of these is the makeup of the pool of incoming engineering students, in terms of both demographics and academic ability.

Census data indicate that the number of 18-year-olds in the population began to decline in 1982, and will continue to fall off until the mid-1990s. It is true that a higher percentage of students have been opting for engineering studies in recent years, but that percentage is variable, so that the overall drop in number of students entering college may become significant for engineering enrollments in the future. An offsetting trend currently is the fact that more women have been entering engineering programs. The percentage of undergraduate female students is now around 15 percent nationwide, but the increase in female enrollments has slowed markedly in the past two years (Engineering Manpower Commission, 1984). Enrollments of Orientals are quite high: 4.2 percent of bachelor's degrees awarded in 1983, for example, went to Asian/Pacific graduates; in California, Orientals accounted for a full 32 percent of undergraduate engineering degrees (Panel on Engineering Graduate Education and Research, 1985). However, enrollments of other minorities, such as blacks and Hispanics, remain low.

Apart from quantities, another limiting factor is the variable ability or preparedness of the student pool. Engineering deans report that SAT scores of entering engineering students are at an all-time high, and have recently surpassed those of liberal arts majors for the first time. Interest in engineering over the past several years has been such that the better-quality schools have had to turn away applicants with strong qualifications, for lack of room. This presents a problem in itself, since it means that potentially talented students are not able to acquire a high-quality engineering education. An interesting corollary of the increased attractiveness of engineering is that the demographics of engineering students have also changed recently: engineering deans and faculty note that many more students are now coming from the suburban middle and upper-middle class.

A different factor that may have implications for engineering supply in the future is that, in general, the level of math and science literacy in the secondary-school population is declining (see, for example, National Commission on Excellence in Education, 1983). Although test scores of current engineering-school entrants are higher than ever, the scores of the overall pool are lower than ever. This trend, if it continues, cannot help affecting the quality of engineering students in the future, particularly as student career choices seem to be strongly affected by shifts in the perceived employment prospects for a given field. The antitechnology sentiment is an underlying current that may once again become overt, as it did in the late 1960s and early 1970s. Because such shifts in perception affect the nature of demand for technological goods and services, they also affect the demand for engineering personnel, and thus indirectly the supply as well. Current engineering students are among the most able in their age cohort. If engineering were to become less popular as a career choice, the drop in quality of applicants could be precipitous. In addition, the fall-off in overall math/science literacy must be viewed against a backdrop of greatly increasing emphasis on math and science in engineering by the year 2000.

Salaries of engineers have been a strong point in attracting students, particularly during the recent inflation/recession cycle. But it is becoming widely recognized that, after the initial five years in industry, engineering salaries tend to flatten out in comparison to other professions (in fact, even in comparison to some skilled workers) (Engineering Manpower Commission, 1983a, 1983b). If there are indeed shortages of engineers, salaries do not reflect that fact. Concern about this and the related issue of quick obsolescence of the engineer may combine to reduce interest in engineering as a career, if the economy continues to improve.

ADAPTABILITY IN THE EDUCATIONAL SYSTEM

The focus of the delivery system for engineers is the engineering educational system, where stresses resulting from changes in the nature and intensity of demand are felt most acutely. Under pressure on the one hand from industry to provide specifically trained graduates, and on the other from students and many professional groups to provide versatile professional education under adverse classroom conditions, engineering schools must be resilient.

Engineering education is subjected to conflicting pressures over the type of preparation it should provide. Essentially three divergent approaches are represented: (1) greater specialization; (2) broader, more general technical education; and (3) the inclusion of far more general content (e.g., liberal arts) in the engineering curriculum.

Arguments For and Against Specialization

The engineering profession has always undergone pressure to strongly specialize engineering education. Industry in particular is often insistent that students do not specialize early enough in their education. This belief tends to be reinforced by engineering faculty within the various disciplines. At the same time, as panel members from industry report, many practicing engineers regret that they did not focus more intensively on their areas of specialization while in school.

However, because of changing technology and demand it is likely that many engineers will find themselves working outside the discipline in which they were educated at some point during their careers. Also, within a given discipline, engineers are likely to find themselves learning and using new skills. This transdisciplinary movement has already occurred on a large-scale several times in the past, and the capacity of engineers to accomplish it successfully has been valuable to industry and to the nation. Thus, educational institutions should be cautious about becoming more compartmentalized and providing more specialized training. Instead, what is needed is a good balance of specialization and breadth of courses in the individual's program as well as in the overall curriculum.

There is a persistent school of thought that argues that, in addition to a broad engineering education, engineers should receive a much more thorough grounding in nontechnical subjects. The rationale here is that exposure to the more traditional elements of a broad, general education would make engineers more well rounded, and thus stronger professionals and better, more flexible engineers.

However it is best accomplished, it seems clear that the uncertainty and unpredictability inherent in the current period argue for a greater, rather than lesser, flexibility in the educational system and its graduates. Some alternatives to greater specialization are emerging that may help to bring about this result.

Alternative Approaches

One useful approach involves emphasis on basic studies—generalized "core" courses for all engineers—in the first two or even three years. This approach is not new—the University of California at Los Angeles was perhaps the first to attempt it, in 1945—but it need not be new to be valid. The basic-studies approach has been successful in the past, and is still being applied by universities today.

Another older practice that still has value is the five-year degree program. Most such programs have been discontinued because of economic competition from four-year programs. Some schools continue to offer the five-year degree as an option, but Dartmouth College is probably alone in maintaining it as a requirement. The extra year affords the opportunity for stronger grounding in the basics (and perhaps in nontechnical subjects) along with greater specialization.

Yet another approach is the "cooperative" program offered by a number of schools, which features several school terms spent working in industry. This approach has the advantage of offsetting the additional expense of a fifth year (through salaries) while affording the student an opportunity to become oriented to work in the "real world" and to make valuable contacts in industry.

Another trend that should be noted is the emergence of the "engineering technology" degree program at several major universities. In addition to providing a broad technical education, these programs train students in drafting and other mechanical skills that are no longer required of engineering school graduates. Many engineering tasks nowadays do not demand a full range of "old" and "new" skills simultaneously. Thus, the engineering technology degree affords companies the advantage of more differentiated staffing.

Another major alternative to greater specialization in engineering schools is afforded by continuing education. Many large industrial corporations now provide some degree of postbaccalaureate training in-house. Many others do not. The expense involved is great (indeed, small companies often cannot afford to offer training at all), but if industry does not feel that schools are turning out a product suitable for its needs, or if experienced engineers are felt to require some "retooling," this is certainly an effective approach. Industry training is not the only avenue of continuing education, however. Schools offer part-time and evening curricula geared to the practicing engineer, particularly in urban areas. This option is often taken solely on the initiative of the individual engineer, perhaps with tuition reimbursement; there is also the possibility of corporations offering part-time daytime schooling as an employee benefit for engineers in certain specializations. Other opportunities for continuing education are offered by professional societies and commercial houses in the form of short courses, seminars, and correspondence courses. Finally, computer-aided instruction at home is becoming increasingly viable with the spread of home computers. The panel expects course-ware offered through this medium to become quite diversified and sophisticated. Thus, there are many opportunities for continuing education, with the majority of them available to any engineer.

THE IMPACT OF TECHNOLOGICAL CHANGE ON EMPLOYMENT

In early nineteenth-century England, as the Industrial Revolution was taking place in that country, sporadic outbursts of sabotage of looms and other steam-powered factory machinery began to occur. The attacks were being made by groups of workmen inspired by the example of Ned Ludd, a possibly mythical Leicestershire weaver. These spontaneous protests by "Luddites" actually delayed the implementation of new technology in certain English industrial centers. In the present-day, the shadow of the Luddite rebellion continues to fall across the concept of automation as one of the potential consequences of technological change.

Potential Impacts on Society

In terms of effects on employment in general, the most significant technological change in the offing is automation—in its modern form, the introduction of computerized systems (whether robotic or not) in the workplace that replace or obviate human workers. One result is technological unemployment or "displacement" of workers. This is a potent political and economic issue. Technology ("mechanization") was blamed by some for joblessness during the Depression, although the actual causes were quite different (Layton, 1973). It is not even certain that large-scale job displacement will now take place. It is likely instead to be a highly dynamic process, with adjustments being made continuously (Office of Technology Assessment, 1984). However, whether or not severe displacement does occur, the panel believes that public perception of it is the key issue. It may well be that, like environmental issues in the late 1960s and early 1970s, concerns about the employment effects of emerging technologies will now be the basis for strong frictions in society. These concerns may do more harm to both human and engineering interests than the environmental issue did and must therefore be addressed explicitly.

The outlook is for substantial displacement of workers over the short run in both the manufacturing and service sectors. The latter is often overlooked; in fact, automation may displace service-sector jobs at a rapid rate. One has only to think of word processing machines with remote printers that greatly increase the output of the individual (and are increasingly used by professionals rather than typists), or large copying machines that auto-feed at high speed, collate, and bind automatically, to begin to envision the scale of effects on the office alone. In any case, it is impossible to predict the amount of displacement that will occur in either the service or manufacturing sector—too many variables are involved. We do not know, for example, how the growth of the service sector is affecting technology, or how technology will respond to new services. The rate of implementation is an unknown, as is the capacity of workers to adapt by any of a number of means. Another important unknown is the degree of resistance that American workers will demonstrate against the implementation of the new technologies.

It is certain that automation will also create jobs at a substantial rate in both the service and manufacturing sectors, although in the service fields these will probably be lower-skilled, low-wage jobs in health services, food services, etc. However, the panel believes that new jobs in this sector will not offset jobs lost or diminished through the introduction of automation.

Taking the long view, the panel concludes that it is possible to be optimistic about the effects of increasing automation on general employment. The economy has historically been very inventive in creating new jobs. Because changes in technology usually bring new industries and increases in demand, they generally alter employment rather than reduce it—although the time-scale can be sufficiently long so that harm to individuals is not prevented. For example, people were displaced from cottage-industry weaving in Europe in the eighteenth century by "automated" looms; but a century later even greater numbers were employed in industrial weaving. Because career mobility is greater today, individuals can more often avoid economic harm. In the United States, people displaced from mining and manufacturing from the 1950s on have tended to enter the burgeoning services sector. It is important, however, not to let such generalizations about trends mask the fact that the negative impact of technological change in many individual lives can still be profound.

The essential point is that, if change is managed well by society, improvement (rather than deterioration) of the quality of life is quite possible. A case in point is the gradual reduction in hours worked per week since the beginning of the Industrial Revolution. The spread of "flex-time" in recent years is perhaps a sign that even the 8-hour workday is beginning to give way to what could become a less-than-40-hour workweek. Labor savings are, after all, one of the major reasons behind the development of automation technologies. There is no reason to believe that their introduction will necessarily have catastrophic effects on society.

Potential Impacts on Engineering Employment

In the context of engineering employment, technological change has impacts not only through automation of manual tasks, but also in the form of new technology and discontinuous change in technology. (The production of a controlled atomic fission reaction might represent the first, while the invention of the transistor is an example of the second.) We have examined a few cases of the emergence of new disciplines in response to demand for a new technology, as well as the response of engineers to the rapid obsolescence of an established technology. In both cases, as long as the change was not too sudden, engineers and the educational system adapted successfully.

The effects of automation on engineering employment are somewhat different, and should be examined separately. There will be considerable displacement of engineers brought about by the implementation, in the manufacturing sector, of computer-aided design and manufacturing systems (Office of Technology Assessment, 1984). It may be that fewer engineers will be required to prepare designs, or to program and monitor robots or flexible manufacturing systems. Much drafting and analysis will be computerized, as will a great deal of documentation. The overall number of engineers employed in this sector may therefore decline. Nevertheless, with reductions of the work force in general, engineers will (in the opinion of the panel) represent a higher percentage of the manufacturing work force than they now do. Manufacturing will become more engineering-intensive.

The outlook for job creation in engineering is possibly better than for production workers. There is now a noticeable call for more manufacturing engineers, a discipline traditionally associated with the "smokestack industries." Contemporary manufacturing engineers will have an important role to play in the application of computers and advanced technology to the manufacturing process. Many engineers will enter the service sector to join consulting firms offering turnkey systems and system start-up and/or operating services.

Perceptions of jobs gained and lost, and of the quality of engineering work in the automated environment, will affect the choices of young people regarding engineering study. Environmental issues influenced students' choice of disciplines as well as the nature and directions of the practice of that discipline. If technological unemployment is to be the next "environmental-type" issue for engineering, similar impacts on choices and directions may occur.

Roles and Responsibility for Intervention

Just as in the case of environmental problems in the 1970s, the government may have to intervene (directly or indirectly) in labor displacement if the application of technology is to proceed smoothly. This seems essential from a pragmatic as well as human-welfare point of view: Society will have to make provisions for severe technological unemployment to avoid a modern recurrence of the Luddite phenomenon. Industry is not and cannot be responsible for the social consequences of decisions taken to ensure survival in the marketplace—although many companies do attempt to take such consequences into account in their business behavior. The formula that is frequently expressed (initially by James Baker, vice-president of General Electric) is "automate, liquidate, or emigrate," with companies threatening to take production offshore if workers and unions will not accept automation. Workers have already tried to prevent both by lawsuits, strikes, and other means; efforts to resist may intensify in the future. Industry and government ought to attempt to find alternatives and solutions in the meantime. There are surely more choices than to automate, liquidate, or emigrate. Carefully thought-out social and technological interventions are needed.

What is the responsibility of the engineering profession in coping with this problem? It should recognize that technological unemployment is a major challenge for the present and the immediate future but also insist that it is not the responsibility of engineers to meet that challenge alone. In fact, it is largely a social problem, one with strong political implications. Engineering professional societies should be aware of the problem, and engineering education should be structured to inculcate in the student the knowledge that engineering is a social enterprise, having social ramifications, and that the innovation and management of complex technical systems often involve considerations of this sort. Here is, in fact, an instance of the value of the kind of "socialization" of engineering education that was urged earlier in the report. In the end, it may be possible for engineers to devise means to automate that accomplish the goal of increased productivity while being sensitive to human interactions and consequences.

SOCIETY'S RESPONSIBILITY TO THE ENGINEERING PROFESSION

Nearly all of the report thus far has emphasized the responsibility of the engineering profession to society in general and the degree of success it has had in meeting those responsibilities. This emphasis is an appropriate one; the profession exists to serve the needs of the larger community. However, it is also important to consider the responsibility that society has to maintain conditions necessary for the continued health of the engineering profession. "Society," in this instance, includes all those entities that benefit from the engineering function —whether they be government, industries, corporations, or individual consumers.

Two primary considerations emerge in this context. The first is the question of whether engineers in general are adequately compensated for their services. An argument can easily be made that compensation of engineers is not commensurate with the value of their contribution to society. The panel believes that the economic productivity of engineers, compared with that of other professionals such as lawyers and financial managers, for example, is high. Yet an informal comparison of incomes shows a great disparity between engineers and those groups. The problem is not at the entry level; beginning engineers earn salaries that are among the highest in any professional grouping (Bureau of Labor Statistics, 1983). It occurs, instead, throughout the middle and later years in the career path—years in which other professionals can expect to reap the rewards (in financial terms) of their experience and seniority. Inadequate compensation for mid-career engineers in academia produces "salary compression," which in turn helps to drive some engineering faculty out of teaching. In industry, it produces a virtual flight of experienced engineers out of technical work and into engineering management, and even into nonengineering fields (Guterl, 1984). This problem is deeply rooted in the nature of our economy and its system of rewards. It is also one that would be extremely difficult (and expensive) to solve. However, a report on the subject of engineers vis-à-vis society would be remiss if it did not at least point out the problem.

The second major issue regarding society's responsibility to engineers relates to the government demand patterns discussed earlier. Although the engineering profession has shown considerable flexibility in responding to past shifts in government demand, the ability of the profession to meet those needs is only one side of the picture. On the other side, considerable hardship is entailed for many engineers in the process—especially for the most experienced engineers. Massive layoffs in defense industries such as aerospace, for example, inevitably put many individuals out of work for long periods of time. Viewing the matter strictly in investment terms, the panel believes that a considerable inefficiency in the use of the nation's technical resources is involved.

Given the rapidity with which government demand can change, and the scale of change involved, it does not seem appropriate to rely completely on the engineering profession to make the great adjustments necessary to meet those demands. The federal government should consider the possibility of providing some form of support network for engineers in industries affected by shifts in program funding. Such a network could include as components retraining programs, compensation packages, and even professional relocation. If similar support is extended to manufacturing workers in changing industries such as the automobile industry, it makes sense to conserve the even more valuable resource embodied in engineering talent, which represents a substantial investment of public funds for engineering education and on-the-job training acquired in government-related development programs.

5

Maintaining Flexibility in an Age of Stress and Rapid Change

Chapter 4 established a general framework for assessing the adequacy of the engineering supply system, from the point of view of both society and the engineering profession. Based on experience up to the present time, a variety of general conclusions were reached about the importance of flexibility and adaptability among engineers and within the disciplines at critical junctures in the nation's industrial/technological development. Basically, the panel finds that the system can respond (and has responded to changing demand for three reasons: (1) the engineering educational system is flexible enough to adapt institutionally and pedagogically to new requirements; (2) students react quickly to economic signals in opting for or against an engineering career and in choosing specific fields of engineering study; and (3) historically, change has seldom occurred more rapidly than individual engineers could adapt. But a number of characteristics of the engineering institutional infrastructure were pointed out as being potential weaknesses in the system, in the face of emerging economic, technological, and social stresses.

The general conclusions set forth earlier on the adequacy and functionality of the system were necessarily tentative, acknowledging the fact that the environment in which the system operates is changing rapidly. What was lacking was some means of understanding more clearly how the system might function under possible future conditions. Accordingly, the panel undertook to project a number of potential scenarios of situations affecting engineering and to use past events as a basis for estimating the response of the engineering manpower supply system. The results of the scenario evaluations are summarized later in this chapter.

HOW WELL IS THE SYSTEM WORKING?

The primary questions to ask in judging the adequacy of the engineering manpower supply system as configured today regard its current responsiveness (in both quantity and quality) and its potential for adapting to future conditions.

Does the Supply Meet the Demand?

In general, the supply of engineers to meet industrial needs and societal goals has proven to be adequate in the past. The response to demand has occurred via three mechanisms. First, engineering schools have accommodated large fluctuations in student throughput; they have also adapted organizationally to pressures for different forms of interdisciplinary engineering study (e.g., environmental engineering). This process has been largely reactive—that is, the institutions tend to be conservative and to make such adjustments only when they are thrust upon them. Consequently, organizational changes and associated changes in curricula have often lagged behind changing demand. Nevertheless, the panel finds that, in general, this element of the system has worked.

Second, individual practitioners have adapted to changing technology in their field by acquiring new knowledge and mastering new skills. Often this is a function of exposure to new technology on the job. In other cases it is a matter of individuals extending their capabilities through some form of continuing education, either within the company or by means of formal course work pursued on their own initiative. When rapid technological change does occur in a particular field (e.g., the introduction of integrated circuits), engineers already working in that field are generally better positioned to keep abreast of those innovations than are (for example) students.

Third, transdisciplinary movement of engineers has occasionally been of major importance in supplying engineers to meet an emerging demand. There are usually enough generic similarities between a new application (spacecraft, for example) and existing ones (e.g., aircraft, submarines, automobiles, and other vehicles) so that specialists in a particular area can transfer their knowledge into the new field with relative ease. The organizational aspects of R&D and production in different fields are sufficiently alike that the difficulty of "plugging in" to a project effort in a different field is minimized for a practicing engineer.

These three mechanisms have enabled demand for engineers to be adequately met, in general, in the present era. There have occasionally been temporary shortages of engineers in specific fields; in recent years this has been the case in electronics and computer engineering. But, thus far, these shortages appear to have been rectified within a reasonable period of time, and before damage was done to either the domestic or international competitive strength of companies entering new areas of technology development.

Is the Quality of the "Product" Adequate?

The initial output of the engineering manpower supply system is, of course, the engineering graduate. Whether this human "product" is adequate to meet the needs of industry is a subject of varying degrees of debate from one industry to another. Clearly, in those fields where change is the most rapid and productivity is the most critical, the pressure for high-quality entry-level engineering employees will be most intense. Currently in the high-tech fields—particularly computers and manufacturing automation—the issue of quality in engineering graduates is being examined closely. The question of quality has essentially three facets: (1) whether engineering graduates come equipped with enough knowledge in their area of specialization; (2) whether, by contrast, graduates possess adequate breadth of multidisciplinary skills; and (3) whether these new employees are sufficiently oriented toward work in the "real world"—that is, whether they write and communicate well and are quick to learn how they fit into the organization and how to work productively on a project team.

Different facets of the contemporary graduate are criticized by different industry groups at different times. Perhaps the only consistent criticism is in the third area, and to some extent the first, in that (based on informal surveys by panel members) new hires often require a considerable period of in-house training before they are capable of functioning productively, confidently, and autonomously in their jobs. Related to this is a criticism by some employers of the large math/science component in the educational background of their new employees. The objection is that the resulting theoretical orientation is impractical for a young engineer on the job in many types of engineering work.

In a less obvious sense, another output of the engineering supply system is the engineers who move into new areas and new disciplines in response to emerging demands. The quality of these "products" is relevant as well. However, because their adequacy has apparently never been a subject of open concern in industry, presumably such engineers are satisfactorily meeting the demands of positions and responsibilities they obtain.

CAN THE SYSTEM FUNCTION UNDER PROJECTED FUTURE CONDITIONS?

Potential Scenarios of the Future

As was pointed out in Chapter 1, one of the main purposes of this report is to ask the engineering profession: "Where have we been; where are we now; and where do we go from here?" Previous chapters have attempted to answer the first two parts of that question. Based on inferences drawn from that analysis, it should be possible to project the future functionality of engineering.

It must be pointed out, however, that to attempt such predictions in a broad sense would be futile. There are too many unknowns, too many variables external to the engineering system, to give any hope of accuracy in assessing the future in general. Since engineering is not a closed system, there can be no satisfactory predictive models. However, it is possible to examine the functioning of engineering under well-defined but hypothetical situations. Therefore, the panel's approach was to propose a set of circumstances ("scenarios") that might occur and that would have an impact on engineering. Their actual likelihood or unlikelihood was not considered to be crucial. The assumption was that it is possible to select isolated events of sufficient range so as to test the capacity of the engineering system for handling stressful change.[1] The scenarios examined were:

1. Continued development toward unmanned factory operation, resulting in the United States regaining world leadership in "smokestack" industries (or, alternatively, losing its competitiveness in manufacturing altogether).
2. Attainment of a recognized capability for commercial utilization of space facilitated by reliable space transportation and permanent in-orbit space manufacturing and laboratory facilities.

---

[1] The selection of scenarios to be examined was based on panel discussion of events that were deemed (a) possible within a roughly 10–15 year time frame and (b) at least potentially capable of exerting severe stress on engineering practice and/or the engineering supply system. Some 15 potential scenarios were considered; 6 were selected for evaluation. Individual panel members were assigned to write one scenario, in which they attempted to project the likely sequence of events and the impact on engineering. Each scenario was then discussed by the full panel.

3. A major new environmental crisis: large-scale contamination of groundwater resources.
4. Widespread adoption of automated teaching via computer.
5. Rapid shift to use of composite materials as a replacement for metals.
6. Sharp fluctuations in the federal budget for defense R&D.

The analysis of the six hypothetical scenarios provided a set of "windows" on the future of the engineering supply system. In each case the panel speculated on what the impact on the engineering community would be, and determined whether (and by what means) the system could cope with the specified circumstances.

Significance of the Scenarios

None of the scenarios appeared to exceed the capacity of the engineering community and the engineering supply system to respond and adapt. This is certainly a positive reflection of the flexibility of the system as currently configured and as demonstrated on several occasions in the recent past. But that is not to say that there would be no pain associated with the response to those conditions; indeed, short-term stresses would in most cases be severe for engineering schools, for companies, and for individual engineers.

It should also be pointed out that the hypothetical scenarios were examined in isolation, as if each one were the only unusual stress being felt by engineering at a given time. In reality, it is likely that two or more such events would be taking place simultaneously, with combined effects that would be much more difficult to predict—and, possibly, to withstand. For example, at the present time there are a number of new technologies whose emergence is not a matter of speculation; they are just arriving or just over the horizon. These include:

- Computer-aided design (CAD), manufacturing (CAM), and engineering (CAE)
- Biotechnology
- Artificial intelligence
- Fusion reactors
- Space-based weapons systems (lasers, particle beams, etc.).

Each of these technologies will have a significant impact on engineering education and practice, particularly when taken collectively. A wide variety of other scenarios can also be projected, most of them no less likely to occur than those that the panel chose to examine. Some of these might be:

- A major worldwide depression
- A strong economic resurgence leading to a "boom" economy
- A critical shortfall of essential materials (e.g., oil)
- A widespread resurgence of antitechnology sentiment
- A quadrupling of the cost of education.

Because of the uncertainty about what events—and how many—might occur that would affect engineering, it cannot be simply assumed that the engineering supply system is well equipped to meet any conceivable future. Each of the scenarios would create stress within the engineering community; even today there are numerous problems of engineering manpower supply, particularly in the area of education. In the context of a discussion of flexibility, it would be well to look specifically at these current stresses.

WHERE ARE THE GREATEST STRESSES APPEARING IN THE SYSTEM?

Under current conditions, a number of points of particular stress can be identified in the engineering community and the engineering supply system. Some of the stress points are perhaps temporary, while others are more long term in their effects; but no attempt is made here to distinguish them on that basis. Instead, they are divided into those that primarily affect the engineering educational system and those that place stress on the engineering community in general.

Educational System Stresses

- The undercapitalization of engineering education; that is, inadequate funding for plant, laboratory equipment, and faculty salaries.
- Overloading of engineering-school classrooms and, conversely, the rejection of some qualified applicants.
- Divergent pressures regarding educational content (more specialization versus generalist technical training versus more liberal arts study).

General Stresses

- Technological obsolescence or displacement of engineers, brought about by both new technology (including automation) and discontinuous change in technology.
- Diminishing pool of 18-year-olds over the next 15 years, resulting in reduced engineering personnel supply.

- Dominance of government demand for engineering goods and services in the marketplace.
- Fluctuating societal attitudes toward engineering and technology, which influence the demand for engineering-intensive products.
- The increased emphasis on factory automation and new manufacturing processes.
- Increased demand for and perceived shortages of engineers trained in information and computer sciences.

6

Conclusions and Recommendations

The following conclusions drawn from the deliberations of the panel are paired with recommendations for action (if any) needed to address that problem or circumstance.

CONCLUSION

In the past, the engineering supply system has responded well to changing societal demand. The engineering institutions have proven to be remarkably adaptable in practice, and individual engineers have generally been flexible in responding to change—although spot shortages and a certain amount of individual hardship have not been entirely avoided. Despite numerous stresses, the system continues to function reasonably well today.

RECOMMENDATION

Because the system is working reasonably well in meeting demand at the present time, no precipitous actions should be taken that would alter its basic functioning under present conditions.

CONCLUSION

Nevertheless, because of the rapidity of technological, market, and social change, the panel cannot be confident that the engineering supply system will continue to be sufficiently adaptive in the future.

RECOMMENDATION

To permit the timely recognition of future problems as they develop, some means of monitoring the functioning of the engineering supply system should be put in place. Based on short-term (e.g., industry recruitment and government research grants) as well as long-term (e.g., trends and projections) data, this approach would provide an engineering-specific complement to the excellent Science Indicators report prepared biennially by the National Science Board. In addition, methods should be explored for increasing the responsiveness of the system at such time as should be required. One approach would be to regularly construct scenarios of events and responses, based on historical case studies and engineering manpower models, in order to test the effectiveness of potential interventions.

CONCLUSION

The system has been able to respond adequately to changing demand largely because: (1) the engineering educational system is diversified and flexible enough to adapt institutionally and pedagogically to new requirements, and (2) students react quickly to economic signals in opting for or against an engineering career and in choosing specific fields of engineering study.

RECOMMENDATION

In order to retain the responsiveness of engineers and of the overall system, engineering schools should not introduce greater specialization into their curricula. Instead, they should continue to emphasize basic skills and interdisciplinary study.

CONCLUSION

The current shortage of faculty makes it difficult for engineering schools to offer a high degree of specialized training while still offering the broad, balanced education necessary for maintaining adaptability in the engineering system.

RECOMMENDATION

Alternate sources of faculty, such as practicing engineers "on loan" from industry, should be developed (although it must be recognized that there are serious disincentives for practicing engineers to participate; nor do all competent engineers make competent teachers). Increased use of teaching assistants and non-Ph.D. faculty would also expand a school's teaching capacity. Perhaps the most exciting potential, however, lies in new ways of teaching. The engineering educational system should utilize educational technology to the fullest in developing alternate methods of instruction. Computer-aided instruction, computer simulations, and the creative use of satellite technology for voice-video-data communications are among the most promising opportunities.

CONCLUSION

Social values and attitudes play an increasingly important role in establishing and altering patterns of demand for engineering-related products.

RECOMMENDATION

Engineering education should be structured to instill in the student the knowledge that engineering is a social enter prise, having social ramifications, and that the innovation and management of complex technical systems involves consideration of social preferences and impacts as well as economic and political realities. Engineers should be trained to view their work in light of anticipated criticism on the basis of social impacts. In addition, the engineering professional societies can be instrumental in informing engineers on these matters and addressing broad political and social issues on behalf of the profession.

CONCLUSION

Both directly and indirectly, the federal government has become a dominant user of engineering goods and services. (Some 15 percent of engineers are employed directly, another 30 percent or more indirectly.) As a result, the panel is concerned about the relative balance in civil and government utilization of these goods and services, and its impact on the strength of the commercial infrastructure. It is also concerned about the ways in which this increasing "public sector" demand affects the structure, content, and orientation of engineering education.

RECOMMENDATION

Some mechanism and methodology should be devised for determining whether (and to what extent) necessary civil applications of engineering goods and services are being compromised through governmental competition. The shifting balance between the market context for engineering and the public context should be monitored by this means. When necessary, government should endeavor to restore a healthy balance through appropriate actions (for example, by improving R&D in support of elements of the commercial infrastructure).

CONCLUSION

The introduction of new techniques and technologies (including all those associated with automation) is likely to create considerable job displacement among workers in both the manufacturing and service sectors. These trends may then generate political and social pressures having strong implications for engineering, as was seen in connection with environmental issues during the 1970s.

RECOMMENDATION

The engineering profession should recognize the seriousness of this issue but should understand that it is also a management problem and a political problem. Mechanisms should be set up to monitor the employment impacts of automation and to identify the points at which political and technological intervention may be useful or necessary. This monitoring should comprise more than just the collection of statistical employment data. It should also include directed studies (perhaps longitudinal) of the economic impacts of technological unemployment on individuals and groups.

CONCLUSION

Current engineering students are generally among the most able in their age cohort, with high ability in science and math as well as strong verbal skills. However, science and math literacy in the overall high school population is declining. It cannot be assumed that engineering schools' students will continue indefinitely to be drawn from the highest ability group. Yet a great increase in emphasis on science and math in engineering work can be expected by the year 2000.

RECOMMENDATION

The engineering profession—in particular, the professional societies—should actively support efforts by government at all levels to enhance the delivery and effectiveness of precollege education in science and mathematics.

References

Armytage, W.H.G. A Social History of Engineering. Cambridge, Mass.: MIT Press, 1961.

American Society for Engineering Education. Goals of Engineering Education (final report). Washington, D.C.: American Society for Engineering Education, 1968.

Bureau of Labor Statistics. National Survey of Professional, Administrative, Technical, and Clerical Pay. Washington, D.C.: BLS, 1982.

Christiansen, D. The issues we avoid. *IEEE Spectrum* 21 (6): 25 , June 1984.

Engineering Manpower Commission. Salaries of engineers in education. Washington, D.C.: AAES, 1983a.

Engineering Manpower Commission. Salaries of engineers in industry. Washington, D.C.: AAES, 1983b.

Engineering Manpower Commission. Engineering Manpower Bulletin, No. 73 , July 1984.

Florman, S.C. Blaming Technology: The Irrational Search for Scapegoats. New York: St. Martins Press, 1981.

Guterl, F. Spectrum/Harris poll: The career. *IEEE Spectrum* 21 (6): 59–63 , June 1984.

Jensen, A.S. The open channel: Does the law recognize software engineers? *Computer* 17 (5): 81–82 , 1984.

Layton, E.T., Jr. The Revolt of the Engineers: Social Responsibility and the American Engineering Profession. Cleveland, Ohio: Case Western Reserve Press, 1971.

Layton, E.T., Jr. (led.). Technology and Social Change in America. New York: Harper & Row, 1973.

National Academy of Engineering. Educational technology in engineering. Washington, D.C.: National Academy Press, 1981.

National Association of State Universities and Land Grant Colleges. Report on the Quality of Engineering Education. Report of the Committee on the Quality of Engineering Education, of the Commission on Education for the Engineering Professions, November 1982.

National Commission on Excellence in Education. A Nation at Risk. A Report to the Nation and the Secretary of Education. April 1983.

Noble, D.F. America by Design: Science, Technology, and the Rise of Corporate Capitalism. New York: Alfred A. Knopf, 1977.

Office of Technology Assessment. Computerized manufacturing automation: Employment, education and the workplace (OTA-CIT-235). Washington, D. C.: Government Printing Office, April 1984.

Panel on Engineering Graduate Education and Research. Engineering Graduate Education and Research. Washington, D.C.: National Academy Press, 1985.

Pursell, C.W., Jr. (ed.) Technology in America: A History of Individuals and Ideas. Cambridge, Mass.: MIT Press, 1981.

Report of the Panel on Infrastructure Diagramming and Modeling. In preparation.

Secretary of State for Industry. Engineering Our Future: Report of the Committee of Inquiry Into the Engineering Profession. London: H.M. Stationery Office, 1980.

Shakertown Conference. National Crises-Severe Engineering Faculty Shortage. Shakertown at Pleasant Hill, Harrodsburg, Ky., 1981.

Yankelovich, D. Science and the public process: Why the gap must close. *Issues in Science and Technology* , Fall 1984.

Bibliography

Armytage, W.H.G. A Social History of Engineering. Cambridge, Mass.: MIT Press, 1961.

Daniels, G.H. The big questions in the history of American technology. *Technology and Culture* 11:1–21 , 1970.

Ferguson, S.F. The American-ness of American technology. *Technology and Culture* 20:3–24 , 1979.

Florman, S.C. The Existential Pleasures of Engineering. New York: St. Martins Press, 1976.

Florman, S.C. Blaming Technology: The Irrational Search for Scapegoats. New York: St Martins Press, 1981.

Jensen, A.S. The open channel: Does the law recognize software engineers? *Computer* 17(5):81–82 , 1984.

Kasson, J.F. Civilizing the Machine: Technology and Republican Values in America, 1776–1900. New York: Penguin Books, 1977.

Kemper, J.D. Engineers and Their Profession (3rd ed.). New York: Holt, Rinehart, & Winston, 1982.

Layton, E.T., Jr. The Revolt of the Engineers: Social Responsibility and the American Engineering Profession. Cleveland, Ohio: Case Western Reserve Press, 1971.

Layton, E.T., Jr. (ed.). Technology and Social Change in America. New York: Harper & Row, 1973.

Merritt, R.H. Engineering in American Society, 1850–1875. Lexington, Ky.: University Press of Kentucky, 1969.

Noble, D.F. America by Design: Science, Technology, and the Rise of Corporate Capitalism. New York: Alfred A. Knopf, 1977.

Pursell, C.W., Jr. History of technology. In P.T. Durbin (ed.), AGuide to the Culture of Science, Technology, and Medicine . New York: Free Press, 1980.

Pursell, C.W., Jr. (ed.) Technology in America: A History of Individuals and Ideas. Cambridge, Mass.: MIT Press, 1981.

Rosenberg, N. Technology and American Economic Growth. New York: Harper & Row, 1972.

Yankelovich, D. Science and the public process: Why the gap must close. *Issues in Science and Technology* , Fall 1984.

Appendix

Engineering in an Increasingly Complex Society

Historical Perspectives on Education, Practice, and Adaptation in American Engineering

A Report Prepared by Arthur L. Donovan

Virginia Polytechnic Institute and State University for the Panel on Engineering Interactions With Society

This report attempts to provide a preliminary yet comprehensive overview of engineering as a social and cultural activity. It draws on historical studies presented at a conference sponsored by the National Research Council: Engineering Interactions With Society: Issues, Challenges, and Responses in the History of Professional Engineering and Engineering Education, held in Washington, D.C., July 19–21, 1983. The report begins by characterizing engineering in three ways: as a distinctive type of knowledge, as a profession, and as a social practice. Three types of adaptation in engineering are then considered through a review of representative cases. The first type involves the interaction of science and engineering, the second the response to technological innovation, the third the influence of institutional factors. The report then examines the relationship between engineering and management and the implications this relationship has for engineering education. The final section of the report reviews selected historical cases of potential crisis in the engineering manpower supply system and the ways in which engineers present their work and their profession to themselves and the general public.

CONTENTS

THE NATURE OF ENGINEERING

Introduction

It would be convenient were we able to begin our investigation of engineering with uncontroversial definitions of what engineering is and what it means to be an engineer. The fact is, however, that engineering encompasses such a complex and highly varied set of activities, and engineers have such a diverse set of skills and interests, that simple definitions are quite incapable of being both comprehensive and useful. Indeed, were we to begin with definitions, we would be answering at the outset, at least by implication, the very questions we have set out to investigate. Therefore, rather than proceeding abstractly and axiomatically, we will approach our subject more tentatively and from several vantage points, always seeking to illuminate its many facets while slowly building a picture of the whole. This is a method of investigation historians find both congenial and informative, but it is not an approach used only by historians. It is a method that those charged with characterizing contemporary engineering also find useful.

The National Science Foundation, which collects statistical information on the education and employment of American engineers, has developed a three-part definition that includes as an engineer anyone who meets two of its three criteria. These criteria, formulated as questions, ask 1) Was the person educated as an engineer? 2) Does the person consider him-or herself an engineer? and 3) Is the person employed in a position classified as an engineering job? These three questions provide a good starting point for an investigation into the nature of engineering, for each directs our attention to a different way of conceiving of the subject.

Asking if a person was educated as an engineer emphasizes the importance of formalized knowledge and knowledge acquisition in modern engineering as well as the role that schools of engineering play in certifying that their graduates are adequately trained to enter the profession. Since control of a specialized body of knowledge is one of the defining features of every profession, the ways in which that knowledge is systematized and transmitted to those wishing to enter the profession is a matter of great importance. While in the past engineers, like other professionals, acquired their characteristic skills through apprenticeship, today formal training in a postsecondary professional school is expected of all beginning engineers. The transmission of formalized knowledge is certainly the main concern of these schools, but we should also be mindful of the ways in which they socialize aspiring engineers in the patterns of thought and conduct appropriate to their profession. Such socialization was clearly a major part of the experience of apprenticeship, and today it remains a large part of what engineers learn during their early years on the job. One particularly fascinating question, but one that is difficult to answer, asks how the responsibility simultaneously to socialize and educate affects the ways in which the central ideas of engineering education are conceptualized and conveyed in engineering schools.

Asking if a person considers him-or herself an engineer directs attention away from questions of public certification and toward the individual's professional self-image. This is not to say that one can simply certify oneself as a professional engineer, for such clearly is not the case. But beyond the educational attainments and memberships in societies that one expects of a professional lie questions of self-description that are of crucial importance to the individual and to the profession of engineering as a whole. What does it mean to conceive of oneself as a professional engineer and how does it influence one's conduct when dealing with members of other professions and with those who are not professionals? And if one moves from a job that requires engineering expertise to one that is essentially managerial, as so many engineers do, in what sense is one still a professional engineer? These are questions of considerable significance to engineers as they fashion their careers and to those who wish to understand better the nature of engineering.

Identifying engineers by referring to the jobs they perform appears to be a direct and uncomplicated way of getting at our central question, yet here, too, the situation is more complex than appears at first sight. There are, of course, certain engineering specialties that are legally defined for purposes of certification. One can also survey engineering employment and identify the various jobs that require certain specialties in engineering. But a closer look at the actual employment decisions and career patterns of those who consider themselves engineers reveals a much greater variety of options and actions than such formal classifications would lead one to expect. Not only do engineers move between specialties, employers in private industry and in the government frequently hire engineers for reasons that have little to do with their particular technical competence. The most interesting question, therefore, is how employers seeking to get a particular job done communicate with engineers attempting to construct rewarding careers. It is the agreements they reach that determine which jobs are to be considered engineering jobs, and seen in this light, it is evident that the list of jobs that fall into this category will vary greatly over time.

Engineering as a Method for Solving Problems

Engineers take pride in "getting the job done." They feel they are particularly well equipped for the tasks they undertake because they bring to them the principles of analysis and problem resolution they learned while studying to become engineers. These principles are commonly referred to as "the engineering method" and they are usually learned in classes devoted to engineering design. Eugene Ferguson, reflecting on his own experience as an engineering student, recalled being taught that "the first thing you do in design is to draw a circle around the system under consideration in order to define the boundaries and control whatever may cross them." He also pointed out that this approach to design, which presumes that the system under examination can be successfully isolated and controlled, was first developed by Italian military engineers in the sixteenth century. Whereas their predecessors had designed fortresses that incorporated whatever advantages were offered by the local landscape, the sixteenth-century Italian engineers argued for a more abstract approach. Favoring a purely geometric and symmetric design to one that embodied local features, they argued that the ideal fortress would be located on an open plain. The surrounding territory was to be stripped of any structures that might give aid to an attacking force, a stipulation that was captured by the pithy phrase of a seventeenth-century French general, "suburbs are fatal to fortresses." Fortress design was still being taught on these principles at West Point as late as 1860, and the more general" engineering method" embodied in this approach to design continued to inform engineering education up to the very recent past.

Ferguson's story may be taken as a challenge to reexamine what we mean when we speak of the engineering method. Can it be that despite the vast expansion of our engineering knowledge since the sixteenth century, we still are using methods of analysis and design introduced over 400 years ago? This is a difficult question, for while on the one hand it is quite clear that in actual practice engineers use many different methods, the idea that there is a method common to all engineering is still a central concept both in engineering education and among those who believe they can identify an approach to problem solving that is distinctive to engineering.

Can the so-called engineering method be defined in a way that enables us to distinguish engineering from other human endeavors? While engineering is a practical activity, so are cooking and child care. And while the engineering method is rational and empirical, so too are the methods used by scientists and judges. We get a bit closer to the specific features of engineering when its method is characterized as reductive. When engineers engage a problem, they sharply delimit the number of parameters examined and focus on those that show some promise of enabling them to control the structure or process in question. While engineering shares with science the search for causal understanding, it differs from science in treating that understanding as a means to control rather than as an end itself. Engineers also differ from scientists in what might be called their propensity for conceptual innovation. Whereas scientists are free to develop new concepts as necessary, while deferring until later questions about the "reality" of the entities they propose, engineers are much more constrained by the need to ensure that the concepts they use in analysis and explanation refer to physical entities and conditions that can be subjected to human control. If this characterization of the engineering method is correct, then this method powerfully influences the determination of which problems are to be considered engineering problems, as well as how those problems are to be analyzed and resolved.

While the above description of the engineering method helps spell out some of the ideas associated with this concept, it remains quite abstract and certainly does not provide a sufficient account of the nature of engineering. Even at the level of method, this generally conceived view of the subject omits all the detail that informs the methods actually used by practicing engineers. It also says nothing about the substantive knowledge that engineers utilize when analyzing and solving problems. As Edward Constant has pointed out, the knowledge engineers find useful can range from the most abstract and general scientific knowledge (one thinks of the Euclidean geometry employed by the Italian fortress builders) to the most specific and context-dependent knowledge acquired by experience (such as the knowledge possessed by the stonecutters who built fortress walls). Engineers spend a great deal of their time acquiring, evaluating, and applying knowledge, whatever its source. In principle they are omnivorous and opportunistic, taking and using information from any source that is able to provide it. In practice, of course, they have developed a variety of means for collecting and screening the flood of information that would otherwise inundate them. Indeed, successful engineers realize there is always a danger that useful channels of information will be closed off, as occurs when the well-known "not invented here" mentality becomes dominant. To understand how engineers function, one therefore must pay attention to the knowledge resources they draw on as well as the methods they employ.

The image of the engineer as an applier of scientific knowledge is in reality dated and quite inappropriate as a characterization of contemporary practice. In the nineteenth century it was thought that the relationships of science, engineering, and society could be captured in a rather simple formula, a crass but representative version of which served as the motto for the Century of Progress World's Fair held in Chicago in 1933: Science Finds, Industry Applies, Man Conforms. But this invocation of a well-worn slogan was at least a generation out of date, for with the rise of the science-based industries at the end of the nineteenth century, most notably the chemical and electrical industries, the relationship between science and engineering became much more complex than it had been. Rather than simply applying the discoveries of science, engineers increasingly had to design and carry out research programs of their own to generate the knowledge of substances and processes that they needed to solve the problems they faced. In the twentieth century, science and technology relate more through interpenetration than through sequential application, but we have not yet developed an understanding of this relationship that will allow us finally to dispense with the slogan that our predecessors found so uplifting.

The realization that in the future engineers would have to generate much of the knowledge they would need naturally brought about a far-ranging examination of the ways in which young men were trained for careers in engineering. The focus of this particular debate has been the issue of creativity. As Michal McMahon has noted, throughout the twentieth century prominent engineering educators have been particularly concerned about sustaining the leading edge or creative sector of engineering. This concern has occupied a central place in the many reports they have produced and remains an issue today.

What is creative engineering? The human capacity to be creative is certainly not something that is entirely the product of formal education, although it can be encouraged or discouraged by the attitudes of teachers and the ideologies of institutions. Thus, within engineering the issue of creativity becomes one of determining what sorts of engineering activities are considered to be of greatest importance and what means are most likely to promote their pursuit. Given the diversity within engineering as a whole, there is no reason to think that any single set of goals or activities will command general assent as being of preeminent importance. And since the word "creative" is a term of high praise in our culture, every active engineer will seek to characterize his work toward the goals he seeks to realize as creative. But we should not avoid the debate over creativity in engineering just because it has a strong tendency to evoke self-serving rhetoric. The issue is too important to ignore, especially because it leads directly into an examination of some of the most important disagreements over values within engineering.

In the present century the debate among engineering educators over creativity has pivoted on the issue of how much and what kind of instruction in scientific subjects should be required of engineering students. Rather than dividing over whether or not engineering students should study science extensively, for all parties agreed they should, the participants in this debate have differed on whether the values of science, and the kinds of knowledge produced under their guidance, are appropriate and fruitful values for engineering. Dugald Jackson, who developed the first cooperative training program in 1907 while serving as head of the electrical engineering department at MIT, believed that the primary responsibility of engineering educators was to prepare their students to serve industry and advance to managerial positions. A thorough grounding in science was needed, but Jackson did not believe that the disinterested and noncommercial values of science were appropriate for engineering and he valued managerial effectiveness over technical creativity. Charles Steinmetz, the legendary General Electric research engineer and a founder and president of the American Institute of Electrical Engineers, opposed Jackson's philosophy of engineering education. He believed the success of modern engineering was a consequence of the progress of empirical science and he was appalled by the degree to which engineering schools continued to stress the acquisition of information rather than the mastery of modern methods of scientific investigation. He argued that while in college, engineering students should study the scientific foundations of engineering and the humanities, leaving until their entry into industry such training in technical practice as they might need. For Steinmetz, the promotion of creativity was the proper goal of education and for engineers the study of basic science was its means.

A generation after Jackson and Steinmetz debated the issue of creativity, William Wickenden again raised Steinmetz's banner in his justly famous 1929 report on engineering education. As McMahon reports, Wickenden concluded that the engineering colleges were so burdened by having to train legions of engineers for the ordinary supervisory and commercial needs of industry that they were largely unfit to train students for the research activities that are also a vital part of engineering. A quarter century after Wickenden's report, Frederick Terman, reflecting on his wartime service as head of the Radar Countermeasures

Laboratory of the Office of Scientific Research and Development, again raised the question Steinmetz had addressed. An engineer himself, Terman concluded that the war had demonstrated the inadequacy of the training engineers received, since most of the major advances in electronics had been made by physicists. Unlike the engineers, the physicists had mastered the basic fundamentals of science while acquiring their advanced degrees, and they were quickly converted into extremely good engineers. The engineers he worked with, while they had functioned extremely well in some capacities, had shown little creativity.

Reflecting on the engineering method, the relationship between science and engineering and the role of creativity in engineering help clarify certain aspects of the overall enterprise called engineering. But consideration of these issues also reveals that no one of them, nor even all of them taken together, provides a basis for a comprehensive understanding of the nature of engineering. Being an engineer involves the use of certain methods and the utilization of certain kinds of knowledge, but it also involves forms of professional association and social practice that cannot be seen as simply derived from its knowledge base. It is to the examination of these other aspects of engineering that we must now turn.

Engineering as a Profession

Engineers have long aspired to the dignity associated with being professional and there can be no doubt that today engineering is one of the largest and most prominent of the professions. What is in doubt is exactly how one should characterize the profession of engineering. One approach is to measure it against the standards of independence, collegiality, and ethical concern that have long been the guiding principles of the older professions of the ministry, the law, and medicine. Another approach is to describe carefully the actual concerns and practices of professional engineers and take these as defining. In fact, both the normative and descriptive approaches are needed, for the powerful urge to professionalize engineering has been motivated both by a desire to elevate the status of the engineer within the larger society and by a commitment to serve the functional needs of engineers as their numbers and specializations have multiplied. These two motivations have created a vitalizing tension within the profession of engineering, a tension that was evident when the first engineering societies were founded and is still present in the profession today.

James Brittain has suggested that one way to step back from the subject of professionalism and bring it into focus is to look at the culture of engineering, using the term culture in the way that anthropologists do. A culture, in this sense, is a system of beliefs embedded in and expressed by a language and related forms of symbolic interaction. These beliefs and their expression provide the context of meaning for those who participate in the culture. This is, of course, a very abstract concept, and cultures only become meaningful to those who are studying them when they are specified by being attached to certain groups of people living at certain times and in particular circumstances. There are, however, two main reasons for thinking about the profession of engineering in cultural terms. The first is that it provides a way of addressing both the prescriptive aspect of professionalization, in which becoming a professional is presented as a way of achieving a higher level of social and personal worth, and the functional aspect, in which being a professional is seen as advantageous in terms of getting on with the work of engineering. The other great advantage of focusing on the culture of engineering is that it enables us to identify what endures, even while changing, in the social organization of engineering. While the culture of engineering is a moving target, as Melvin Kranzberg has put it, the processes by which young engineers are acculturated and socialized into the profession are still of great importance. It is those processes, and the goals they are intended to serve, that we need to understand.

One way to spell out the content of a culture is to look at its dominant images. Larry Lankton has suggested that until quite recently the engineer was perceived as a Lone Ranger of the technical world. When help was needed he rode in and fired off a few silver bullets, saving the day for the virtuous and serving the public good. The image is arresting if fanciful, and in some ways it captures the character of men like John Jervis, arguably the foremost American civil engineer in the nineteenth century. In addition to being an accomplished engineer, Jervis laid great stress on his personal integrity and independence. Since the construction of civil works inevitably involves politics, Jervis cultivated an image that would enable him to speak with authority in public debates. He realized that technical expertise, while necessary, was not enough and he therefore stressed his autonomy as a professional and his personal independence. When consulting on the construction of the Croton Aqueduct, for example, Jervis specified that a certain type of mortar be used. The commissioners insisted that a different mortar be used and Jervis, convinced that their decision was technically unsound, told them that while he acknowledged their authority to overrule him, if they did so he would withdraw from the project. Jervis's reputation was such that he prevailed.

The American Society of Civil Engineers (ASCE), founded in 1852 and the first American engineering society, was dedicated to defending and advancing the image of the professional engineer represented by Jervis. But the civil engineers' emphasis on personal autonomy was only possible in fields in which engineers were in fact independent. So long as engineers operated primarily as consultants bound to their employers only by job-specific contracts, they could see themselves as professionals in practice in the same sense as practicing lawyers and physicians. But in fields such as mining, in which many engineers were employed by particular companies for extended periods and served in managerial as well as technical capacities, the image of the professional as an independent agent was much less sustainable. Indeed, one need only glance at the early history of the American Institute of Mining and Metallurgical Engineers, the second of the four "founder" societies in engineering to be established, to see that for some engineers, unstinting loyalty to the company of a sort one would expect of someone in a managerial position was entirely compatible with the development of a sense of professionalism. (See Edwin T. Layton, Jr., *The Revolt of the Engineers*, Cleveland, Ohio: Case Western Reserve Press, 1971.)

The other two "founder" societies established in the nineteenth century, the American Society of Mechanical Engineers (ASME) and the American Institute of Electrical Engineers (AIEE), attempted to strike a balance between the influence of business and the independence of technical expertise. Since mechanical engineering developed out of a machine-shop culture that had been evolving throughout the century, its leadership came largely from those with extensive experience in the metal-working industries. A tension soon arose between those who wished to ensure that the shop culture of mechanical engineering would continue to dominate the profession and those who looked to a more formally organized and transmitted school culture as the proper foundation for mechanical engineering. The AIEE, which served engineers in one of the new industries spawned by scientific discovery, set high technical standards for membership, but it also acknowledged the dominant position of the new electrical companies and accepted a high degree of business leadership within the society.

This brief review of the various conceptions of professionalism that informed the four founder engineering societies clearly reveals that the culture of engineering is highly diverse. Indeed, each of the major societies appears to embody a distinct subculture, each of which attempts to reconcile the competing interests of greatest importance to that branch of engineering. If arrayed along a continuum, the various engineering societies would be bounded at one end by scientific societies, which are preeminently devoted to the discovery and interpretation of natural knowledge, and at the other end by industrial trade associations. The tensions that shaped these societies at the time of their founding, tensions such as those between practical experience and theoretical understanding and between individual autonomy and loyalty to one's employer, are still present in engineering and continue to challenge those responsible for engineering education and the affairs of engineering societies.

Engineering as Social Practice

The practice of engineering, like the engineering profession, also can be analyzed in cultural terms. In an earlier historical period, when there were many fewer engineers and engineering specialties than there are today, it was not unreasonable to think that all engineers shared a single set of professional values and followed careers that conformed to a common pattern. But today this belief in a common culture of practice, while still informing certain aspects of engineering education and professional organization, does not provide an adequate basis for understanding the actual work experiences of contemporary engineers. For well over a century the growth and diversification of engineering has been driven by the development of new technological systems that require both new types of knowledge and new forms of social organization. Edward Constant described one consequence of the general shift in engineering practice these developments have brought about when he observed that today, "virtually all engineering . . . is done in complex organizations, either in industry, in government, in education, or somewhere else. There are very few solitary engineers. Engineers have a reputation for being casually antisocial and yet virtually everything they do requires fairly intense social interaction." To acquire a more detailed understanding of contemporary engineering practice, however, we must move from this level of generality down to the study of the particular subcultures of engineering practice that taken together make up today's culture of engineering.

As James Brittain has pointed out, two recent books, both of which were bestsellers, provide extended and revealing accounts of the culture of engineering practice in two different industries. The first of these, Kurt Vonnegut's novel *Player Piano*, was published in 1952. It is, of course, a work of fiction and must be interpreted with care, but it is also an insightful description of certain attitudes and patterns of behavior that the author observed while working for General Electric in Schenectady. One need not share Vonnegut's views about the events he describes nor consider him a particularly skillful novelist to appreciate the sharpness of vision he brings to his study of the subculture of engineering at General Electric.

Vonnegut is centrally concerned with understanding how corporations go about indoctrinating new engineers. How are these young people persuaded that they should see themselves as part of the corporate team and how do they come to internalize and make their own the company's view of the significance of their work? This is obviously a question of great importance, for all social organizations must develop ways of ensuring that those who function within them will demonstrate considerable concern for and allegiance to the goals the organization is attempting to realize. We should therefore not object to Vonnegut's concentration on this facet of engineering practice, but should rather be prepared to examine the issue he raises and the conflicts engineers experience as important aspects of the culture of engineering practice. This, of course, is not the place to undertake an extended discussion of Vonnegut's treatment of these questions. It is worth pointing out, however, that the dramatic action of the book culminates in the annual corporate camp meeting at which the company engineers renew their adherence to the values which they, as company employees, live by. And as Brittain reminds us, General Electric did in fact have a camp on Association Island where its engineers and managers went to learn, play, and revitalize their commitment to the corporation.

The second book Brittain summarized, Tracy Kidder's *Soul of a New Machine,* is an outstanding example of contemporary reporting. Kidder immersed himself in the daily life of a group of engineers who were given the task of creating an entirely new computer, his purpose being to describe dispassionately yet vividly the character of computer engineering as it is actually practiced today. He tells us that competing teams, the Hardy Boys and the Micro Kids, were formed and that those who wished to sign on to a team had to undergo a rite of initiation. The project leaders told the teams they were going to build what they could get away with. As with the formation of teams, problems were analyzed in binary terms—answers were right or wrong, decisions were good or bad. Dedication had to be complete, project members being told at one point that they were expected to ruin their health for the company. Kidder tells his story with great skill and provides a compelling account of heroic obsession within the computer industry. One can, of course, argue that the book is unrepresentative, but it has been widely praised as accurate by those familiar with this kind of work. At the very least it shares with Vonnegut's novel the virtue of directing attention to aspects of engineering practice that are seldom analyzed.

While those who wish to understand the nature of engineering should welcome the appearance of books such as Vonnegut's and Kidder's, putting them in the hands of engineering students may, from one point of view, prove counterproductive. Engineering education, like other forms of professional education, must inspire as well as inform, and it is reasonable to ask whether engineering students ought to have their attention directed to the aspects of engineering practice highlighted by Vonnegut and Kidder. Indeed, there is some evidence that engineering students have a strong sense of self-preservation on this score. David Hounshell reports that he has used Kidder's book in a course for engineering students and that 75 percent of the students consider it the book they like least. The book they most like, David McCullough's excellent history of the building of the Brooklyn Bridge, can also be read as a story of heroic obsession, but in this case there is a central figure who overcomes the many difficulties he encounters and leaves as a monument to his triumph a massive structure whose beauty and economy is still being celebrated. The extent to which engineering educators are responsible for tempering the unrealistic expectations of their students is, of course, a pedagogic question of considerable importance.

Not all engineers work for corporations, and Martin Reuss has provided an informative account of the ways in which the subculture of engineering in a federal agency helped shape the practice of one of its foremost figures. Reuss has studied the career of Andrew Humphreys, an army officer who rose to become chief of the Corps of Engineers and was also one of America's most prominent hydraulic engineers during the nineteenth century. Humphreys is remembered for his influential study of the hydraulics of the Mississippi River. He brought to that project the engineering training he had received at West Point and his belief that science and engineering should be supported by, and should in turn serve, the state. It is interesting to note that both his approach to hydraulic engineering and his views on the relations between science, engineering, and the state were derived from Continental rather than British sources. The Corps of Engineers was committed to serving the public good, although whether it in fact always did so would be seriously questioned in the twentieth century. Humphreys respected this noncommercial mandate and vigorously resisted those who sought to reduce the corps to an agency that merely contracted with engineers in private practice for such services as it needed. He is therefore a suitable representative of the subculture of public engineering for the era in which he lived.

Humphreys certainly recognized that politics influences engineering decisions, but he insisted that his own recommendations were informed by engineering considerations, not political interests. As chief of engineers, he used data and ideas first published in his famous Mississippi delta report to defend his positions on controversial engineering issues. As a result, his disputes with various professional engineers took on the character of a referendum on the report itself, and since Humphreys had tied the corps so closely to the conclusions of the report, judgments concerning his work as an engineer also were taken as judgments of the entire corps.

The best known of these disputes was between Humphreys and James B. Eads, whose most famous work is the great bridge that still bears his name and spans the Mississippi at Saint Louis. Eads was convinced that building jetties at the mouth of the Mississippi would provide the channel depth needed for navigation; Humphreys argued vigorously against the jetties project. The outcome of the dispute was determined by the interplay of egos, alliances, and politics as much as it was by dispassionate analysis of competing engineering theories. Congressmen, most without any engineering education or expertise, debated highly technical questions and called upon government and, increasingly, nongovernment experts to testify before their committees. Newspaper coverage of this dispute was so extensive that Reuss considers it the first technical engineering debate that became a national issue.

Humphreys brought to this engagement his status as chief of engineers and his reputation as co-author of an important report dealing with river engineering. He was proud of his work as a hydraulic engineer and proud of his West Point education. Eads, with little more than a grammar school education, brought to the contest not only his own considerable talent, but also the brash entrepreneurial drive of an independent professional determined to win public business for private contractors. Humphreys was outraged as Congress intervened in ways that politicized the entire debate. As it turned out, Eads was right on the technical issue, but this obviously does not provide grounds for concluding that in general private engineers are smarter than public engineers. This case is important for our purposes because of the light it throws on the character of federal engineering in the nineteenth cen tury and, more particularly, on the ways in which the Continental model of engineering in the service of the state informed one subculture in the history of American engineering.

An awareness of the cultural differences within engineering may be of some help to those who are seeking to open the professions to under-represented segments of the American population. Indeed there are some indications that certain cultural groups are differentially attracted to the various subcultures within engineering. Whether or not women should be seen as a distinct cultural group in this sense is open to question, but as Robert Saunders has observed, they have entered the profession in great numbers recently and they are apparently gaining acceptance as a result of their abilities as engineers. He believes that within 20 years women will constitute 40 percent of the engineering work force. Asian-Americans have also moved strongly into engineering and those who have recently emigrated from Southeast Asia are continuing this tradition. The story is very different with blacks and chicanos, however, for the serious efforts that have been made to attract them to careers in engineering have been largely disappointing. As engineers and others continue the struggle to meet the nation's commitment to affirmative action to achieve equal opportunity, they may find it worthwhile to attempt a more precise fit between the subcultures of engineering and the cultural characteristics of the peoples they seek to attract.

PATTERNS OF ADAPTATION

Science and Engineering

Most attempts to describe the complex relations between science and engineering are compromised at the outset by partisan preconceptions. Scientists, eager to demonstrate the utility of their increasingly expensive research, emphasize useful "spinoffs," and indeed they can cite enough cases to make the argument plausible. Engineers can reply, however, that the same discoveries, or equally satisfactory solutions to the problems solved by these discoveries, well might have been found more quickly and at considerably less expense had the problems been attacked directly. Even within the realm of science itself engineers can point to the crucial role of technology. Melvin Kranzberg, developing an argument first advanced by the late Derek Price, has suggested that much of modern science, especially in those fields that depend on elaborate instrumentation, should be seen as applied, or perhaps theo rized, technology. All such arguments assert that either science or engineering is the more fundamental of the two activities, the other being essentially dependent. It should be evident, however, that while this either/or interpretation of the relationship between science and technology may enable us to understand certain special cases, it is completely incapable of providing an account of how these two human enterprises relate in general. And since we have no adequate general theory of their relationship, it seems best to return to the study of cases, but without bringing to that study prior partisan commitments.

It would be easier to distinguish between science and engineering if they did not have so much in common. Perhaps the best way to highlight their differences is to see them as separate cultures, in the sense of having different systems of values for the determination of significance. The use of a common language is no bar to the formation of distinct subcultures within a nation or of different cultures among nations. Science and engineering shared a common mathematical and methodological language, but they differ culturally in the meanings they attached to the uses of that language. Their distinctive systems of meaning are not, of course, completely self-enclosed, for intercultural communication is both necessary and commonplace. What we do find when we turn to history, however, is that in some cases this communication between the cultures of science and engineering has been relatively easily effected and has worked to the mutual benefit of both parties, whereas in other cases it has led to confrontation and breakdown. Jeffrey Sturchio's description of the American chemical community's response to the crisis created by the cutoff of German synthetic organic chemicals during World War I is a case study in the successful mediation of the differences between science and engineering, whereas James Hansen's account of the troubled career of the aeronautical engineer Max Munk at the Langley Research Station can be read as a case in which science and engineering failed to adapt to one another. Both stories should be instructive for those concerned with making the best use of the resources of both science and engineering.

Although the United States had a well-developed chemical industry before World War I, the world market in the important area of synthetic organic chemicals was dominated by German chemical firms. The German firms had several important advantages, including an outstanding tradition of chemical research, the ability to secure product patents in America, and extremely low U.S. tariffs. These advantages enabled them to maintain a near monopoly, even in the United States, on such chemicals as coal-tar dyes and intermediates, certain medicinals, and synthetic nitrogen compounds and other synthetics. When war broke out in 1914, the Germans threatened to embargo all exports of synthetic organic chemicals and the British began to blockade German shipping. The crisis these actions created in the American chemical community generated a response that was so well grounded and successful that 10 years later U.S. production of synthetic organic chemicals had been increased tenfold and long-term control of the market in these chemicals was firmly in the hands of the U.S. industry. Here then is a story of the successful harnessing of scientific and engineering resources in a time of national crisis.

It was crucial to the success of this response that the challenge was perceived to be national and not just a problem for a particular industry. While any downturn in the chemical industry would have had implications for the economy as a whole, synthetic organics were essential for the production of explosives and certain medicines, and hence they were judged to be crucial for national defense. Federal officials therefore assumed responsibility for coordinating the response to this shortage. The recommendations of a committee of prominent chemists convened by the New York section of the American Chemical Society were accepted, and a protective tariff was imposed to encourage investment in the research laboratories and production facilities that would be needed to make America independent in the area of synthetic organics. By the time the United States entered the war in April 1917, the government had, by contracting with several leading chemical companies, built several major plants to produce these scarce chemicals.

Once at war with Germany, the U.S. government provided even more support for the chemical industry. German dominance of the U. S. market depended heavily on patent protection for specific products such as aspirin. This was a type of technical knowledge that could be immediately and directly expropriated. It should be noted, however, that this is not always the case, for technical knowledge frequently resides in the experiential skills of small groups of practitioners, a form of knowledge that cannot be easily expropriated. In 1917 the United States sequestered German property in America, including over 4,500 chemical patents, and assigned it to the newly established Office of the Alien Property Custodian for management. Two years later the Chemical Foundation was established and the licensing of the sequestered chemical patents was assigned to it. The Chemical Foundation used the fees it received to provide public relations and research support for the American chemical industry.

After the war had ended the chemical community and the federal government continued to cooperate. A protective tariff was maintained while the industry adapted to peacetime markets and positioned itself to maintain control of the synthetic organic chemical sector. By the mid-1920s the industry had developed the institutional structures it still has today, including its close ties with the government and research universities.

Jeffrey Sturchio has emphasized a number of interesting aspects of this case. For example, the number of professional chemical engineers grew very rapidly during this period: "From less than 900 students in 1910, there were on the order of 5,000 students in chemical engineering programs in the U.S. during the late 1920s." It was also an era in which the agenda for chemical research in the universities was set largely by the needs of industry: "In the 1920s those departments, such as Columbia University's Department of Chemical Engineering and the University of Illinois' Department of Chemistry, that had very close ties with industry through consultancies, fellowships and other mechanisms, found themselves prospering in ways that other departments did not."

This was also the era in which chemical engineering achieved a position of distinction and prominence within American higher education. The crisis in synthetic organic chemicals, and the rapid professional and institutional growth it helped stimulate, occurred just as Arthur D. Little's famous concept of unit operations was gaining acceptance as the distinctive method of chemical engineering. Chemical engineers, Little asserted, should analyze chemical processes into the unit actions, such as pulverizing, mixing, and heating, that are the elementary steps in the production of industrial chemicals. Chemical engineering quickly became identified with the use of this method and in this way distinguished itself from scientific chemistry. But the separation of the two subcultures of chemistry was not complete, for both were closely allied with industry.

As Sturchio also points out, the American Institute of Chemical Engineers tried to match the number of chemical engineers being trained to the needs of industry, but without notable success. Setting up programs to train chemical engineers inevitably involved a considerable lag time and there was no way to predict whether demand would still be high when the schools were finally operating at full capacity. In fact, early demand estimates included a considerable backlog, so that an overshoot developed fairly rapidly, and by the mid-1920s there were more chemical engineers available than the industry could employ. Since the demand for engineers was strongly linked to overall economic activity, downturns in the economy in the 1920s exacerbated the difficulties of matching supply and demand.

James Hansen's study of Max Munk at Langley presents us with a case in which the differences between diverse subcultures were so great that they could not be bridged either by dedication to a common purpose or forebearance. Munk, a prominent aeronautical engineer born in 1890, was educated in his native Germany. Highly gifted in both mathematics and science, he received two doctoral degrees, one in engineering at the Hanover Polytechnic Institute and one in physics at Goettingen. His mentor at Goettingen, Ludwig Prandtl, considered Munk his most talented student, even when compared with Prandtl's more famous pupil Theodore von Karman. Munk came to America shortly after World War I and began working for the National Advisory Committee for Aeronautics (NACA). For six years he was stationed in Washington, where he designed experimental equipment and worked on theoretical problems for the Langley Aeronautical Laboratory at Hampton, Virginia. In 1926 he was appointed chief of the Aeronautical Division and moved to Langley. Within a year the engineers in his division were in full revolt and Munk was forced to resign.

Why were the engineers at Langley and this highly talented individual unable to work together? The case is complex yet revealing. The easiest way to explain the rupture would be to emphasize Munk's inability to conform to the established expectations and patterns of conduct of the Langley engineers. But we should be wary about going too far in this direction, for Munk was a very talented engineer, and since engineers are supposed to be receptive to all forms of useful knowledge, whatever their sources, it will not flatter the Langley engineers to say that they were unable to make use of Munk's undeniable talent. Hansen has therefore looked more deeply into the nature and origin of Munk's attitudes and behavior, which appeared so eccentric in the Langley setting. For our purposes, this perception of eccentricity can be characterized as a cultural dissonance that arose when Munk's approach to engineering came into conflict with the practice of the Langley engineers.

Munk's ideas about the nature and values of engineering were those of the German university system in which he had been educated. Broadly stated, the highest value within this system, at least with regard to natural knowledge, was attached to theoretical knowledge of the sort exemplified by the exact physical sciences. Mathematicization, theoretical innovation, and individual creativity, values normally associated with the pursuit of pure science, were, in the German universities, also the governing values in engineering. Munk, like Charles Steinmetz, whose views on the primacy of creativity were discussed above, had internalized these ideals and attempted to honor them in his work at Langley. He had also internalized the hierarchical social attitudes of the German university system, where each department was under the strict control of a single professor. The conflict at Langley can thus be seen as rooted in fundamentally opposed views on the nature of engineering knowledge and the nature of engineering as a social practice.

Munk's behavior at Langley certainly was unusual by American standards. He considered himself the absolute master of the division he directed and, like a German university professor, expected to set the research goals for all members of the division and receive primary credit for all the division's accomplishments. He offended the junior engineers at Langley by treating them like German graduate students. They were obliged to attend a theoretical seminar that he conducted in a way they found rude and condescending. They also considered his supervision of experiments vague and overbearing and found his analysis of problems obtuse and excessively mathematical. This was not simply a confrontation between theory and practice, however, for many of the Langley engineers were well trained mathematically and they both acknowledged Munk's personal ability and shared with him the general goal of developing better aeronautical theories. When they found they could not work with Munk, they attempted to work around him. When that tactic failed, all the section heads of the division resigned. When Munk refused reassignment, he was forced to resign, at which point the section heads resumed their positions.

The causes and consequences of the revolt against Munk are still the subject of debate. What can be said, descriptively, is that the in-group at Langley, the so-called "NACA nuts," found the cultural dissonance created by his working there too great and Munk himself an unacceptable eccentric, this despite the fact that Munk was the best classical theorist ever to work at Langley. In cultural terms, the case appears to demonstrate that within engineering the attitudes and values of specific subcultures are frequently of greater importance than the more general values, honored by all engineers as well as others, of seeking reliable knowledge and looking for practical solutions. It is these more specific patterns of belief and behavior that have the greatest bearing on the engineer's ability to make use of knowledge from diverse sources and to adapt to changing circumstances within engineering.

Responding to Innovation

Innovation, the development of new products and processes and their introduction into standard practice, is, like creativity, an aspect of engineering that all engineers consider important, and yet the actual experience of adapting to innovation can be very upsetting. Engineers, like other people working in relatively stable jobs, learn to use familiar materials and follow established procedures as they go about their business. They are, of course, expected to suggest ways of improving how their work gets done, and by and large engineers welcome the kinds of incremental innovations that can be fairly readily integrated into established patterns of work. But the appearance of a major innovation can have a revolutionary effect on the way certain engineering tasks are accomplished. When this happens, the existing organization of work and the knowledge and skills employed are subject to reexamination, and the adaptations required to accommodate truly novel devices and procedures frequently create great stress within specific subcultures of engineering and great disruption in individual careers. The invention of the transistor and its introduction into electrical engineering was a case of this latter type. Robert Friedel has studied this case in detail and his investigation helps us understand the kinds of difficulties and dislocations that a major innovation can cause in a specialized field of engineering.

The invention and utilization of the transistor was one of three major innovations in electronics since World War II, the other two being the development of integrated circuits and of microprocessors. The transistor was invented by three physicists working at Bell Laboratories, Walter Brattain, William Schockley, and John Bardeen, but as the laboratory authorization for the research that led to this invention reveals, the research program was closely linked to the perceived needs of the Bell system. New switching devices and other components were required to handle increasing telephone traffic, and the research directors at Bell Labs had reason to think that a fundamentally new approach to these problems might be fruitful. In July 1945 the three inventors of the transistor were therefore authorized to undertake research in solid-state physics while concentrating on "the fundamental investigation of conductors, semiconductors, dielectric insulators and other electric and magnetic materials." After listing the specific materials to be investigated, the Research Authorization stated why this research was considered important: "Communication apparatus is dependent upon these materials for most of its functional properties. The research carried out under this case has as its purpose the obtaining of new knowledge that can be used in the development of completely new and improved components and apparatus elements for communication systems." The prescience of this authorization is extraordinary.

Two and one half years after this authorization was issued, the Bell research team had invented the point-contact transistor, an amplifier that was the first clearly operable solid-state analog to the vacuum triode. With the advantage of hindsight, which, as Melvin Kranzberg reminds us, cannot readily be converted into 20/20 foresight, we can now see that this was an invention of great consequence, but in December 1947 its implications were hardly evident. Bell Labs, with characteristic conservatism, let six months pass before publicly announcing the invention. And while the first point-contact transistor was the direct progenitor of the junction and field-effect transistors developed subsequently, it was itself a transitional device that had very limited immediate utility.

Students of innovation are fairly familiar with the stages by which the transistor entered the world of electrical engineering, but many of the engineers involved found the process highly disruptive. While economists talk abstractly about the substitution of technology B for technology A, historians know that the introduction of a radically new device or procedure almost always alters both what is produced and the process of production in ways that are entirely unanticipated. Those involved in the process of integrating fundamental innovations into existing systems of design and production begin by treating them as direct analogs of certain elements in well-known systems. They begin, as the economists suppose, by attempting a direct substitution of the new for the old, but then, as previously unnoticed properties of the new devices are discovered and the analogy with the element of the old system begins to break down, the implications of the innovation become apparent. This, in rather general terms, is the way major innovations progressively transform technological systems, finally rendering useless previously established ways of operating. And this is the kind of effect that the introduction of the transistor had within electrical engineering.

The electronics community was at the outset fairly well prepared to welcome certain features of the new transistors. The intense development of radio and electronic engineering during World War II had led to an enormous reduction in the size and weight of tubes and circuits and had made possible the creation of such devices as the proximity fuse. In fact, the degree of miniaturization achieved during the war was proportionally greater than that brought about by the use of transistors. Electronic engineers were therefore keenly aware of the advantages of small size, low power requirements, and ruggedness, and they welcomed the transistor because they believed it promised great improvement on each of these points. Thinking they could substitute transistors for existing vacuum tube rectifiers and amplifiers, they believed the new devices could be utilized without any fundamental reconceptualization of their design criteria.

In practice the transistor could not simply be substituted for vacuum tube rectifiers and amplifiers. Not only were transistors fundamentally different devices having peculiar secondary characteristics that had to be taken into account, the first transistors were not themselves reliable or well understood. It soon became apparent, therefore, that the use of transistors would require both a reconfiguration of fundamental circuitry at a time when electronic technology had reached a high level of maturity and complexity and a great deal of development of the transistors themselves. Friedel cites one engineer who considered this a rather unrewarding prospect: "The transistor in 1949 did not seem like anything very revolutionary to me. It just seemed like another one of those crummy jobs that required one heck of a lot of overtime and a lot of guff from my wife. It wasn't exciting, not exciting at all. My job in the factory was to turn someone else's dream into salable hardware."

Despite these difficulties, however, a transistor revolution was effected in electronics. The pivotal year was 1952, by which time Western Electric was manufacturing transistors in earnest. Although they still cost at least eight times more than comparable vacuum tubes, transistors found a market in miniaturized hearing aids. While this demand encouraged further developments in transistorized circuitry, the big push for new developments in electronics in the 1950s came, as it had during World War II, from the military. The market for consumer electronics, having been relegated to second place by American manufacturers, was taken over largely by European and Japanese firms, while U.S. firms concentrated on designing and producing electronics for the space and arms races and on the miniaturization of computers.

Friedel draws several challenging conclusions from his study of the introduction of the transistor. One is that in engineering, practitioners frequently have to incorporate new knowledge and slough off old knowledge and skills that are no longer useful. This in itself is a fairly commonplace observation, but when the implications of the new knowledge are as revolutionary as they were in the transistor case, the process may be quite stressful. It would be pleasant if all knowledge change occurred in such small steps that the practicing engineer could stay abreast by reading a few articles and taking the occasional short course. In fact, changes of such magnitude may occur that the central problems the engineer confronts have to be reconceptualized and the design principles brought to bear have to be radically reconstructed. One may be obliged to accept complexity in components that one previously sought to simplify, or machines that were always regarded as single purpose may have to be designed to be multifunctional.

Changes of this magnitude can seriously disrupt established patterns of thought and practice.

Friedel further suggests that the introduction of innovations having the revolutionary potential of the transistor may not be completed until a new generation of engineers has replaced the older generation that worked with the displaced technology. This suggestion is not based on any presumption that the new generation of engineers will be in any absolute sense better educated than its predecessors. Their chief advantage will consist of not having internalized the patterns of thought and behavior associated with the older technology. Harsh as this suggestion appears in human terms, it may simply reflect at the level of the engineering work force a pattern that is visible at the corporate level. Jordan Baruch has pointed out that not one of the major vacuum tube manufacturers succeeded in becoming a major supplier of solid-state devices, and Edward Constant has also noted that no airplane engine company that built piston engines also built jet engines of its own accord.

Can anything be done to prevent this displacement of active and useful engineers in mid-career? One option, that of somehow blocking the introduction of innovations with revolutionary implications, clearly is unacceptable. It appears, therefore, that engineers need to be encouraged to prepare to adapt to such changes when they occur. Erich Bloch has noted that there are certain industries that are forced through a cycle of technological obsolescence every few years and that these are the industries that have learned how to survive in the face of rapid change. He has also suggested that leaders in industries that are challenged by rapid technological change must accept responsibility for ensuring that their employees, including their engineers, are prepared to adapt. It must be assumed that everyone who once acquired the knowledge and skills required to do his or her job is also capable of acquiring new knowledge and skills when that becomes necessary. Employers, and especially large corporations, should make the new knowledge accessible to their employees, partly because they have an obligation to do so and partly because they know best what is needed. The resources available in universities may prove useful, of course, since many universities already support extensive programs for continuing education. But universities are broadly concerned with the production and transmission of knowledge and hence can be rather slow in responding to the educational needs created by innovations in industry. The general lesson that Friedel's study teaches is that certain types of technical changes oblige us to intervene in the "natural" process of generational displacement within engineering. Since we can no longer accept or afford widespread loss of employment as a consequence of technological innovation, we must uncouple generations of knowledge from human generations and ensure that every engineer acquires as many generations of knowledge as he or she needs to have a full and productive career.

Institutional Imperatives

The complex organizations within which most engineers work are purposeful institutions, and the strategies and mission statements formulated to express their purposes help define the contextual constraints that shape the practice of engineering. From the setting of research agendas through the development of new processes and products to the evaluation of results achieved, the goals of the institutions within which engineers work provide the primary criteria for determining the success or failure of the effort expended. To grasp how institutional goals impinge on the practice of engineering, we must focus on the strategies that govern specific industries and agencies, for the more abstract goals of "profitability" and "public service," which are honored by all corporations and public bodies, are too general to be informative. We must, in other words, turn again to case studies. Stuart Leslie has looked closely at the ways in which Charles Kettering shaped his work as an engineer to the specific corporate strategies of the firms he worked for, and his case study illustrates how this adaptation takes place in the private sector. Thomas Carroll has studied the development of rockets in a publicly funded laboratory, a case in which the changing research orientation of the lab imperiled but did not terminate the line of investigation that eventually proved to be the most fruitful. Both cases demonstrate that institutional imperatives play a central role in engineering.

Charles Kettering was born in 1876 and received a degree in electrical engineering from Ohio State University. He then went to work at the National Cash Register Company, where he quickly learned that the key to success lay in coupling his talents as an engineer to the needs and opportunities of greatest concern to the company he was working for. These needs and opportunities, he discovered, were most evident to the people responsible for marketing the company's products. Thus, rather than focus his attention on new ideas suggested by recent developments in electrical science, he did his best to provide technical solutions to problems identified by the corporation's sales force. As Kettering later recalled, "I didn't hang around much with the other inventors or the executive fellows. I lived with the sales gang. They had some real notion of what people wanted." His personal strategy was highly successful. By the time Kettering left the National Cash Register Company in 1909, he had helped transform the cash register, as Leslie puts it, "from a defensive measure against weak-willed cashiers . . . into a powerful tool of management planning."

Kettering left the National Cash Register Company so that he could go into business by himself designing and supplying electrical accessories for automobiles. Again he was spectacularly successful, his most famous new product being the first commercially successful electric self-starter. While others before Kettering had developed various self-starting systems, he concentrated on fitting his system to cars already being produced and on making it reliable. By realizing these goals, he made the world of motoring available to new groups of consumers, most notably women, and greatly expanded the market for his new product. In 1916 he merged his Delco Company into General Motors, and a few years later he was put in charge of a research group in GM. His assignment was to study the long-range problems of the industry, especially those that might be of greatest concern to GM's production divisions. It was a task for which he was well prepared both by experience and attitude, and again he succeeded brilliantly.

Managing the Research Laboratory was a difficult assignment, for it was an anomalous unit within General Motors. Unlike the production divisions, it was not a profit-making unit. GM president Alfred Sloan was sharply aware of that fact. Shortly after becoming head of GM Sloan cautioned Kettering that "the more tangible [the] result[s] we get from [the research lab], the stronger its position will be." But the research lab could not simply focus on solving existing technical problems of production, for it was also responsible for reaching out beyond the range of existing products in an attempt to anticipate where the market would go, and this entailed the possibility of making wrong guesses. To satisfy this second expectation, Kettering developed a variety of techniques for identifying what kinds of new devices might sell and he then used these educated guesses when setting the research agenda for the lab. He recognized the importance of risk-taking and told his colleagues in the lab that, "you are always too late with the development if you are so slow that people demand it before you, yourself, recognize it. The Research Department should have foreseen what was necessary and had it ready to a point where people never knew they wanted it until it was made available to them."

Serving the production divisions proved to be a considerably more demanding task that Kettering had at first realized it would be. This point was driven home by his failure to convince the production engineers that they should adopt a radically new air-cooled engine developed by the lab. Having been reminded that the products of the lab were of no value to the company unless they were acceptable to the engineers working in the production divisions, Kettering thereafter devoted a great deal of time and attention to what he called the research lab's internal market. He sought to ensure, largely through personal diplomacy, that ideas proposed by the lab were acceptable within the company before they were developed further for the external market of car buyers. Producibility and marketability were the twin criteria by which the lab's efforts were to be evaluated, and under Kettering's guidance it served General Motors well.

Thomas Carroll's study of the development of solid-propellant rocket boosters at the Jet Propulsion Laboratory illustrates how changing institutional commitments can shape engineering efforts that later turn out to be unexpectedly successful. While engineering teams must respond to the changing concerns of the institutions in which they work, they also develop a certain momentum of their own. The relationship between the larger institution and the practice of engineering is thus a kind of dance in which the institution leads while allowing its partner a certain degree of freedom. Maintaining such a relationship serves both parties well where there are no formulas for success, for in such cases a certain measure of tolerance appears to be practical wisdom. In the case of solid-propellant rockets, it was individual conviction and group momentum that led to successful development, not the sustained commitment of the sponsoring institution. But had that institution not provided some, if limited, resources for those who believed in solid propellants, their conviction and momentum alone could not have resulted in success.

Prior to World War II the Guggenheim Aeronautical Laboratory at the California Institute of Technology (GALCIT) sponsored a small research project using liquid-fueled rockets for high-altitude sounding of the atmosphere. When the war began, the laboratory took on the task of designing rocket motors to provide jet-assisted takeoff for airplanes (JATO), and the sounding rocket project was set aside. After extensive theoretical and developmental work, the engineers assigned to the JATO project perfected a solid propellant that could be cast in heavy containers. These rockets, which had short burning times, were then produced and used in great number during the war. They also served as the paradigm for further development of solid propellants.

Reports of German development of the V-2 rocket led to a dramatic redirection of the Cal Tech research facility in 1944. GALCIT was transformed into the Jet Propulsion Laboratory (JPL), its mission being to develop "long-range rocket missiles and ram jets." The rockets called for by this mission had to have long burning times and the JPL researchers therefore again turned their attention to liquid-fueled rockets. The JATO project was not abandoned, however, although work on solid propellants was demoted to a position of secondary importance within the lab. By 1947 the leaders of the JATO program were convinced that the new designs and new materials they had developed now made it possible to construct a long-burning, lightly cased, solid-propellant rocket. However, when they urged that their research be supported under the long-range missile program, their claims were greeted with skepticism and they were dismissed as eccentric.

From 1948 until 1950 the advocates of solid propellants used the resources made available to them to develop and test a series of large rockets. Their enthusiasm had gotten ahead of their research, however, and the tests were cancelled following the twelfth consecutive explosion of one of these multithousand-pound rockets. During this period JPL had become even more deeply committed to liquid-fueled rockets, as well as to the development of other aspects of long-range rocketry, such as guidance systems, that it had taken on. Although the cause of the solid-propellant rocket explosions had been discovered just as the series of tests was being cancelled, the cost of developing these rockets further had come to exceed the commitment of the institution that had fathered them. At that point, as Carroll puts it, JPL declared the solid-propellant rocket an orphan. Although banished from the estate, the orphan did not perish. Rather, it migrated, in the person of the leading scientists involved in the project, to the Thiokol Corporation, a patron that recognized its true qualities and launched it into a flourishing career.

Thomas Carroll's study of solid propellants at JPL illustrates that setting agendas for engineering research is a gamble. Everyone wants to support "creative engineering," to use Steinmetz's phrase, but the line between creativity and eccentricity is frequently hard to discern. While successful innovation is a common goal for those involved in engineering research, the factors that lead to successful innovation are hard to identify. Martin Reuss has suggested that since institutional goals have such a great influence on the setting of research agendas, "the burden today is on the managers rather than the educators to provide the opportunities for engineers to do . . . innovative work." Yet corporations and agencies must exercise some control, or their research and develop ment groups might concentrate only on "technically sweet" projects while slighting those that seem most likely to serve the needs of the larger institution. What is clear, therefore, is that considerations of institutional mission and strategy are and will remain primary in the planning of engineering research and development.

ENGINEERING AND MANAGEMENT

The Engineer and the Marketplace

There is an intriguing tension in engineering between the public image of the engineer and the reality of the normal career pattern within engineering. The engineer is in the popular mind a skillful manipulator of physical substances and the organizer of vast technical enterprises. Engineers strive for efficiency in the sense of using energy and materials in the least wasteful way possible and they seek to maximize utilities by satisfying the material needs of humanity while making the most efficient use possible of the resources provided by nature. Conceived in this way, engineering can be readily distinguished from management, for this latter activity involves primarily the organizing, motivating, and supervising of people. The goals of management are entirely those of the institution in which the manager works, while engineers are primarily, although certainly not exclusively, concerned with material efficiency and utility. Engineers deal with things, managers deal with people.

But this distinction, while clear, is artificial, for in fact engineering and management blend into one another in ways that make it very difficult to disentangle them. Consider the case of Kettering, described above. Thomas Hughes has asked if Kettering was in fact an engineer when he was serving as director of the GM Research Laboratory. Certainly his responsibilities at that stage in his career were primarily those of a manager. Aaron Gellman notes that a similar problem arose in two university programs in transportation engineering. Social and management issues were introduced to such an extent, he reports, "that people who were turned on by engineering stopped coming," and these programs in transportation engineering were subsequently transformed into programs in transportation management. But at exactly what point did this move from engineering to management occur?

While distinguishing between engineering and management remains a difficult analytical problem, it appears that in practice few engineers today find the relationship between these two enterprises especially troubling. This is so because engineers have by and large accepted and made their own the fundamental values of the institutions they work for. In America in the twentieth century the institutional values of greatest importance to engineers have been those of corporate capitalism, and considerations of cost and profit are as central to contemporary engineering as is knowledge of the properties of the physical world. This historical coupling of the central cultural values of engineering and management has been immensely successful, and there is little reason to think that the ability of this extraordinary cultural compound to motivate and inform the design and production of new technologies is nearing exhaustion. But the fact that these two sets of values have been yoked together in practice does not provide grounds for believing that they are essentially identical. Indeed, the persistence of the distinction between engineering and management in both public opinion and everyday language, as well as the tensions that occasionally arise between these two activities, indicate that while engineers in practice may successfully compromise the differences between the value systems of these two enterprises, their differences still need to be made explicit.

Earlier in this century it was widely believed that the leading management problems created by the rise of industrial society could be successfully reduced to engineering problems in the sense that they could be adequately analyzed and solved in terms of material and energy efficiencies. While today this approach to social engineering has few public adherents, the arguments advanced in its behalf serve as a useful reminder that the conception of the relationship between engineering and management has varied through time. The early champions of scientific management and of the political movement called technocracy were practicing engineers who believed that the wastefulness of competitive capitalism and the inefficiencies of interest-group politics could be eliminated by treating all problems of management, both public and private, as engineering problems. If this were done, the struggle between labor and management over control of the workplace could be resolved through a scientific determination of the organization of work. Similarly, the shortages and unequal distribution of essential goods and services, problems that are a commonplace in the political economy of corporate capitalism, would be eliminated through rationally organized production and distribution. The primacy of the market would be replaced by the primacy of reason as represented by science and applied by engineers. No one today needs to be reminded that this version of social engineering did not prevail. But it did express a clear notion of the difference between engineering and management, one that was accepted by many engineers.

While the technocrats in particular were emphasizing the differences between the ways that engineers and capitalists solve problems, most engineers were happily following career tracks that carried them from technical work to managerial responsibility. Indeed, it has long been a commonplace among engineers that the road to success leads to management. Over 50 years ago William Wickenden, a giant in the history of engineering education, reported that a survey of engineers who graduated between 1884 and 1924 revealed that roughly two thirds had become managers within 15 years of leaving college. Wickenden applauded this finding, for he realized it is almost always necessary for an engineer to leave the engineering of materials and enter the engineering of people in order to become very successful financially and socially. The movement of engineers into management has continued unabated, and many observers believe that if American industry is to hold its own in international trade, the number of engineers in top management must be dramatically increased.

Today the importance of good engineering in the design and production of consumer products is being reemphasized, while at the same time the importance of linking the work of engineers to the marketplace is also being stressed. Striking the right balance between engineering considerations and marketing possibilities is central to the art of management, and as David Hounshell's study of the competition between Ford and GM in the 1920s demonstrates, that balance can shift very suddenly. And as Neil Wasserman's study of recent changes in American Telephone and Telegraph also indicates, corporations that in the past have been organized on functional principles derived from their engineering practice may suddenly find themselves compelled to reorganize to give primacy to market considerations. In general then, those engineers who plan to move into management, and they constitute a majority, must be prepared to accept the primacy of managerial values, which are the values of the marketplace, even when this does some violence to the values they acquired while training to become engineers.

In 1908 Henry Ford introduced the Model T and five years later he revolutionized automobile production by introducing the moving assembly line for major subassemblies and final chassis assembly. Until the mid-1920s Ford's classic car—black, spare, cheap, and reliable—and the technology with which it was produced represented engineering efficiency and utility to most Americans. While the Model T did in fact undergo some improvement and modification over the years, it was never radically altered, and therein lay its vulnerability in the marketplace. In 1922 Ford manufactured over two million Model Ts, held a market share of 55 percent, and reaped huge profits; four years later its market share was down to 30 percent and drastic action was called for.

The alternative to Ford's "best engineering solution" conception of the automobile was developed by GM under the presidency of Alfred Sloan. Rather than offering the public a single automobile containing a compromise of all those features one looks for in a car, Sloan developed a family of cars, or as he put it, a car for every purse and purpose, and in doing so he consciously sought to capitalize on the public's willingness to pay for comfort and conspicuous consumption. By 1925 Sloan's "trade-up" marketing strategy had penetrated each of GM's product lines and the annual model change was introduced. As an engineer, Sloan realized how technically demanding and resource wasteful such a strategy was, but it made great sense as a way to sell cars and Ford was obliged to conform.

In 1927 Ford announced it would end production of the Model T and introduce a new car, the Model A. The changeover, as Hounshell points out, was a disastrous episode in the history of the Ford Motor Company. The costs and difficulties entailed were grossly underestimated and the time required to complete the changeover greatly exceeded original projections. The Chevrolet Division of GM continued to press the market strategy Ford was attempting to copy, and by the time the Model A was becoming profitable, it was already being seriously challenged by a newer and better GM product. Success in automobile marketing had come to depend on continuous innovation in product design and GM had developed the management organization required to implement this strategy. While both companies employed highly competent engineers, Ford was hobbled by an arbitrary and unsystematic management organization that reflected the prejudices and conduct of its founder and owner. The model change fiasco made the inadequacies of Ford's management painfully evident.

Perhaps Sloan should be seen as an "ideal type" of the engineer who became a successful manager. There can be no doubt that he devoted skills he acquired as an engineer to serving General Motors, but whether the decisions he made while president were strongly influenced by his background in engineering remains problematic. Consider the question of installing safety glass, for instance. To do so would certainly have been advantageous from the point of view of safety engineering, as Sloan realized. But as Hounshell points out, Sloan consid ered questions of profitability paramount. Indeed, he put the point as a matter of stark inevitability, not personal choice, saying "I regret that we have to be so selfish that we must consider our profit position before we do, perhaps, the safety of those who use our products, but it cannot be otherwise." Here is an acceptance of the "naturalness" of the market that is as absolute as the naturalness of the physical world engineers must deal with in their technical work. It thus appears that to succeed as managers, engineers must be prepared to accept the naturalness of the laws that govern the social world in the same way that as students they were taught to accept the natural laws of the physical world. But this should not be surprising, for such beliefs are essential components of the systems of ideas that unify and sustain the cultures of engineering and management.

The ways in which market considerations come to dominate technical considerations when engineers work in competitive industries is also illustrated by Neil Wasserman's study of the American Telephone and Telegraph Company. Wasserman uses the phrase functional atomism to describe the managerial system employed by AT&T from the late 1880s until the end of the 1970s. In this system the organization of managerial units paralleled the organization of the engineering functions performed by the various components of the system. When fully developed, the system assigned research to Bell Labs, equipment manufacture to Western Electric, long distance service to a separate division, local service to regional operating companies, and so forth. This organization of management responsibilities was particularly effective at a time when the telephone company was operating as a regulated monopoly having as its primary goal the development of a universal system of voice communication. Shielded from market competition, it was able to control and phase in new technology to ensure functional compatibility throughout the system.

Today the situation in which the telephone companies operate is dramatically different. AT&T has been broken up, competing technologies are being introduced by aggressive entrepreneurs, new kinds of services are being marketed, and universal telephone service, having been largely achieved, is no longer a suitable corporate goal. In response to these changes, Wasserman reports, AT&T decided even before faced with divestiture to move from a managerial system based on functional atomism to one based on market organization. As in the automobile case, the end of what was essentially a monopoly forced those with managerial responsibility to respond to the new range of choices available to consumers and rely less on products that represented engineering solutions to technical problems.

In the final analysis, the relationship between engineering and management, like that between engineering and science, should be seen as one of vitalizing tension, not destructive antagonism. Managers, like generals preparing to fight the last rather than the next war, are inclined to overstress the importance of the market. But as recent events have demonstrated, most noticeably what Simon Ramo has called America's "technological slip," good engineering frequently has its own commercial value, and its absence can be just as damaging as an emphasis on engineering values alone. Managers need not be engineers, and not all engineers have the desire and talent needed to be effective managers, but in today's world engineers and managers must work together if they are both to succeed.

Educating Engineers

Education is inevitably a pragmatic activity. Its ideals are quite properly set very high, but in practice it consists of a series of compromises forced upon both students and teachers by the very real limits of time and resources. But accommodations between ideals and reality are not the only compromises required in the construction of educational programs, for even within the realm of ideals there are competing claims as to what ought to be the goals of education. This is certainly the case in engineering education. Of course, there is no fixed law that says all engineering degree programs must attempt to realize the same goals, and in such a large and heterogeneous field one would expect considerable diversity. But the engineering community does exercise an unusual degree of oversight in the area of professional education and it is not inclined to allow a great number of conceptions of the purpose of engineering education to flourish simultaneously in the name of tolerance. Choices must be made, and one way of laying out those choices is to review some of the recently proposed goals for engineering education.

Samuel Florman has argued with great eloquence and considerable force that if engineers truly wish to be considered professionals, then they ought to structure their professional schools accordingly. Lawyers and physicians do not begin their professional training until they have completed their undergraduate education, usually in the liberal arts. The purpose of their college educations is to make them informed and sensitive individuals and citizens, people who have studied what it means to lead "the examined life" and who can articulate what their rights and duties are as members of the communities in which they live. These lessons can, of course, be learned elsewhere as well, but by treating the bachelor's degree as the professional entry degree, engi neers show a willingness to give less attention to the development of individual talent and culture than do members of what have traditionally been called the liberal professions. The result, Florman charges, is that most graduates of engineering programs today are "nothing more than high school graduates who have taken a lot of technical courses," a view that Florman is not alone in holding. To solve this problem, the professional education of engineering students should be extended so that they can take more courses in the liberal arts. Florman realizes that making such a change would be difficult and that only a minority of engineers consider it desirable. He has, however, highlighted a real limitation in existing programs in engineering education and he has effectively articulated one set of ideals by which the performance of those programs can be evaluated.

An ideal of engineering education diametrically opposed to Florman's informs the many programs in engineering technology that have been set up recently. As Melvin Kranzberg has pointed out, these new associate and bachelor degree programs in engineering technology represent both a conservative reaction to the growing emphasis on basic science and engineering design in mainline engineering programs and an innovation that increases the flexibility and hence resilience of the engineering profession. Engineering technology programs focus on mastery of engineering practices, such as surveying, shop practice, and drafting, that in the past were standard components of an engineering student's training. Students who possess these skills are capable of holding entry-level jobs as engineers, but without a more extensive grounding in mathematics, science, and design, they are ill equipped to proceed on to higher levels of engineering practice. While it is still too early to say how these new programs will be integrated into the profession as a whole, it seems likely that practitioners trained in engineering technology will serve in technical and professional support roles comparable to those filled by nurses and medical technicians in the practice of medicine. The development of this "second stream" in engineering education may thus serve the profession well, but the ideal of education it represents is not one that most educators would find appropriate for schools that seek to provide a more comprehensive introduction to engineering.

It appears that for the foreseeable future the bachelor's degree will continue to serve as the professional entry degree for engineers. For the great majority of engineering schools, curricula leading to the bachelor's degree require two types of courses, the first being scientific/technical courses, which occupy roughly 85 percent of the student's credit hours, the second being social/humanistic courses, which occupy the remainder of the student's required credits. While there is considerable pressure from both sides to change the ratio between these two types of courses, the case made for increasing the technical content at the expense of the social/humanistic seems, on balance, no more compelling than the arguments made for increasing the liberal arts content at the expense of the technical. Everyone involved in engineering education can point to at least one essential area in which graduates of existing programs are woefully ill prepared, but in the absence of a general willingness to increase the number of years of study required, it appears unlikely there will be any reduction in the number of courses required in either of the two branches of engineering education. Engineering curricula are already extremely crowded and highly constrained, and it is impossible to add new requirements to existing four-year programs.

There is considerable evidence that engineering educators have been continuously adapting the content of the scientific/technical side of the engineering curricula to the needs of the profession. In recent years the number of required technical courses associated with specific engineering specialties has been reduced while the number of required courses in basic science and mathematics has been increased. One effect of these changes has been to expand the area of commonality between the specialized curricula within engineering, a move that has been facilitated by an increase in the number of cross-specialty appointments being made in engineering departments. Thus, while engineering students still enroll in specific curricula, such as civil engineering and chemical engineering, they in fact take a great many courses in common and are not nearly as specialized as it might appear. They are thus well equipped to move between specialties as the needs of industry and their own interests may require. When doing so, they of course must acquire the specialized knowledge required for the jobs they take, but this kind of training is increasingly being made available by employers, who recognize that they cannot ask the engineering schools to send them young men and women fully trained in the specialties needed by industry. In light of these trends, it seems unlikely that we will see a proliferation of new specialized curricula at the undergraduate level in schools of engineering during the coming decades, and it may be that throughout the profession of engineering the distinctions between the various branches of engineering will become less important as practicing engineers learn to take full advantage of the flexibility implicit in the content of their professional education.

The purpose of the social/humanistic requirements of engineering curricula, and their adaptation to the changing character of the profes sion, has occasioned a great deal of discussion. The original reason for requiring engineering students to study European history and English literature was to give them a taste of liberal education, the hallmark of a person of culture and of the professional. Then, as engineering became more closely tied to employment in large business enterprises, nontechnical education came to play a more immediately instrumental role in the education of engineers who would become managers. An understanding of the principles of market economics became important, as did an ability to write clear expository prose. This tension between liberal education, which can be understood as the study of the cultural classics for the purposes of self-development, and instrumental education, conceived of as the acquisition of concepts and skills that will prove useful in one's employment, has remained a source of both confusion and vitality in engineering education, as well as in American higher education more generally, throughout the present century. Given the limitations of time, should engineering students take a course in Shakespeare or one in technical writing? Only those who can afford to ignore the constraints imposed by reality are free to say they should take both.

Following World War II leading engineering educators realized that the context of engineering was changing rapidly and that even from an instrumental point of view, the content of the social/humanistic side of engineering curricula needed to be reconstructed. The Engineers Council for Professional Development, the accrediting agency for engineering education, began this process by stipulating that every engineering student was to take at least one course in the social sciences or humanities during each term of study. The next step, according to Melvin Kranzberg, involved the American Society for Engineering Education which, with the aid of a grant from the Carnegie Corporation of New York, prepared a report on general education in engineering. Attempts to implement the recommendations of this report led to rapid expansion of the humanities and social science departments at many engineering schools and the development of what Kranzberg calls the "contextual approach" to the presentation of these subjects to engineering students. This approach begins with problems and situations of immediate concern to engineers and then uses the insights and methods available in the social sciences and humanities to clarify them and make them intelligible and thus manageable. The instrumental character of this enterprise is made clear by Kranzberg's formulation of its guiding purpose: "I look forward to the day when the humanities and social sciences will serve as tools for the engineers, just as much as his computer and engineering handbook."

Not all engineering educators are free to subscribe to a completely instrumentalist view of the social/humanistic requirements. Students attending engineering schools located in comprehensive colleges and universities are normally required to satisfy college or universitywide distribution requirements that cannot be met by taking courses that are primarily instrumental. But even in those schools where engineering educators have effective control over the curricular requirements set for their students, the question of what subjects students ought to study in the required social/humanities courses continues to occasion lively debate. And indeed it should, for identifying and articulating the social and cultural factors of greatest importance to contemporary engineering, and the ways in which engineers experience them and respond to them, is a challenging task. It is one thing to declare one's acceptance of the "contextual approach"; it is much more difficult to make clear just what that context is.

There is good reason for believing that the context of engineering, and indeed the context of management as well, has been radically and permanently altered in the past two decades. The heroic or Lone Ranger image of the engineer has been largely replaced by the image of the engineer as a morally ambiguous actor in society. Where once we celebrated the extension of control over nature and the expansive use of natural resources, today we worry that such activities might be signs of myopic pride and may be contributing to insupportable insult to the environment. Within the industrial order, federal regulation has been extended into the processes of design, production, distribution, and utilization in ways that previously would have been considered unthinkable. Engineering and management decisions on the design and production of automobiles, the mining of coal, the use and disposal of chemicals, and in any number of other areas must now be made with constant and detailed reference to governmental specifications and regulations. The context of engineering has, in other words, become exceedingly complex. Prior to 1960 one could make sense of most engineering activities, at least in the private sector, by referring to the engineering imperative to maximize physical efficiency and utility and to the corporate imperative to maximize profits. Today, focusing on these criteria alone would result in a fatal neglect of many additional considerations that have become a central part of decision making in engineering and management. As many American industrial leaders realize, the technology-forcing and technology-limiting consequences of public policy play an increasingly important role in the nation's economy. While the importance of that role will fluctuate as the political winds shift, there seems to be little likelihood that engineers and managers will in their day-to-day activities ever again enjoy the degree of freedom from public scrutiny and control that they did through the period of the Cold War.

The first reaction to the new level of contextual complexity is, quite naturally, to insist that engineering students spend much more time studying the social sciences and the humanities, but as we saw above, there is little likelihood that more time will be made available for these subjects. A second and more promising response is to say that an instrumental approach to the social/humanistic component of engineering education now compels us to recast instruction in these subjects so that engineering students will be able better to understand the fundamental concerns and claims that lie behind the new public attitudes and policies. If in the public mind engineering now appears morally ambiguous, then the reasons for that attitude and their implications for engineers can be examined in courses on ethics for engineers. And if the conduct and consequences of industrial activity are now to be closely regulated, then the reasons for doing so and the consequences entailed can be examined in courses on engineering and public policy. Such courses, if treated with the seriousness they deserve, can help engineering students think their way through the challenges that will be thrown in their way by those outside the profession. Rather than being driven to sectarian self-justification, they will be prepared to manage the complexities they encounter ultimately to satisfy the highest goal of both engineering and management by getting the job done.

ENGINEERING AND SOCIAL CHANGE

Resilience in Times of Crisis

Engineers find their jobs in a highly differentiated labor market that is both extremely free and highly responsive to change. One can speak of an engineering manpower system, but to do so is to aggregate and rationalize in the abstract a dispersed series of negotiations and contracts arrived at freely and independently between employers and employees. The constraints within the system are imposed, on the one hand, by the total number of potential employees available and the special skills they bring to the marketplace and, on the other hand, by the needs, both total and in terms of specific skill requirements, of potential employers. When it appears the system is malfunctioning, it may be because of an oversupply in the total number of engineers seeking work or in an oversupply in one or more specialties, or, con versely, it may be that the total demand, or the demand for one or more specialties, exceeds the supply. In practice, of course, crises within the system first become evident as shortages or oversupplies within certain fields of engineering. The response to crises involving a shortage in a certain specialty can take two forms. The number of engineers available in the undersupplied specialty can be increased either by increasing the number of beginning engineers trained in the specialty, or engineers trained in other specialties can be hired to do the work required. How the engineering manpower system has in the past responded to shortages, and whether or not individual engineers have successfully migrated between specialties, is thus an empirical question that can be answered, at least in part, by the study of appropriate historical cases. The two cases described below indicate that in fact the engineering manpower system has been surprisingly resilient in times of crises, primarily because large numbers of engineers have in practice been highly flexible in terms of their ability to move successfully between specialties.

Edward Constant, who is studying the early history of petroleum engineering, has been impressed by the extent to which engineers have moved into new and undersupplied specialties from adjacent areas of science and engineering. In 1920 there were only two university programs in America for the training for petroleum engineers. A survey of those who prior to 1920 were doing the kind of work that came to be associated with petroleum engineering reveals that only 9 percent were trained in this field. Of the 147 practicing engineers in the survey who had degrees, over one quarter had received degrees in geology, another quarter had degrees in mining engineering, and the remaining half held degrees in chemistry or other fields of engineering. As new academic programs in petroleum engineering were developed, this cross-flow between specialties naturally diminished. A sample of 180 degree-holding petroleum engineers in practice between 1930 and 1960 indicates that roughly 44 percent had degrees in petroleum engineering, while 21 percent had degrees in geology and 35 percent held degrees in other fields of engineering, with mining engineering accounting for only 5 percent.

Constant's data suggest an interpretation that he believes is misleading. Perhaps the petroleum engineering case is an example of the emergence of a specialty, and once the field has reached maturity, in the sense of having its own degree programs, the flow of engineers into that specialty from other fields will decline to relative insignificance. But as both Constant and Jeffrey Sturchio point out, the assumption that mature specialties operate as closed systems in the engineering man power system is not supported by the evidence of history. If the fluctuations of the system were predictable, these specialty subsystems well might establish an internal equilibrium, for they are strongly inclined in this direction. But in fact the demand for engineers, both in the aggregate and within separate specialties, is affected by so many factors, and the lag time involved in recruiting and training new specialists is so long, that in times of crises a considerable cross-flow between specialties is evident even in mature fields. For instance, in the area of petroleum engineering the 1973 oil embargo, an event that certainly evaded prediction, created a decade-long sharp increase in the demand for petroleum engineers. While this heightened demand led to increased enrollments in degree programs in petroleum engineering, it was satisfied in the short run primarily by an influx of engineers who moved into petroleum engineering from related areas in science and technology. The resilience of the overall engineering manpower system was again demonstrated, and it seems reasonable to attribute that resilience at least in part to the openness of the specialty subsystems of which it is composed.

Alex Roland has drawn similar conclusions from his study of NASA's Apollo program. Driven by a fear of military vulnerability and a desire to demonstrate national power, the lunar-landing program involved engineering on a national scale and threatened to create intense stresses in the engineering manpower system. This threat was relieved in part by certain organizational choices made within NASA. Rather than developing the Apollo program on the Army arsenal model, in which almost all the engineering work is done in-house, NASA adopted the Air Force contracting system and consistently spent 90 to 95 percent of its budget on contracts with industrial suppliers of products and services. Having made this choice, NASA then hired a cadre of its own engineers to plan, supervise, and coordinate its contracts and operations. The engineers hired by NASA came from a variety of specialties, again illustrating the predominance of cross-flow in periods of high demand, and many of its engineers and managers were detailed to NASA from the military services. As a result, NASA never suffered from a shortage of qualified engineers. Although Roland has not studied the flow of engineering manpower in the corporations that contracted with NASA, his impression, shared by others familiar with this story, is that there, too, cross-flow between specialties was the key to meeting the sudden increase in demand for aerospace engineers.

The sudden expansion of NASA associated with the Apollo program was followed by an equally unanticipated sudden decline. NASA managers, seduced by the technical sweetness of the devices they were creating and lulled into believing there was a boundless national commitment to the exploration of space, planned for continued high levels of growth within the agency, but as early as 1963, long before the first lunar landing in 1969, political support for post-Apollo projects had begun to wane. Since 1965 NASA's budget has been steadily declining and it is now less than the military space budget. While this retreat from the space frontier has received a great deal of highly charged publicity, it appears that during the period of decline the engineers in NASA and in the corporations with which it has contracted have either successfully returned to the jobs they held before the Apollo program or have taken the experience they gained while on that project and applied it elsewhere. Thus while both the expansion and contraction of the Apollo program had the potential for creating a crisis in the engineering manpower system, that system in fact exhibited a surprising degree of resilience in responding to the stresses placed upon it.

The realization that the engineering manpower system possesses a high degree of resilience has important implications for engineering education. Because we are incapable of predicting with a useful degree of accuracy future shifts in the demand for engineers, and because the response times of universities are so slow in comparison with those of the marketplace for engineering labor, attempts to tie the content of engineering education closely to the needs of industry have been of little use in anticipating or responding to short-term stresses in the engineering manpower system. Indeed, attempts to forge a tight link between engineering curricula and specific employment opportunities have probably done more harm than good from the point of view of individual flexibility and the resilience of the system, for they have emphasized specialization at an early stage of education and have thereby reduced the breadth of understanding that in fact facilitates movement between specialties.

The character of the engineering research carried on in universities appears to have a considerable bearing on the flexibility of the engineers trained within them. The most effective link between college-and university-based engineers and the markets served by engineers appears to lie in the realm of research. While it is relatively easy to insure that research and development activities carried on within a corporation are market responsive, such is not the case in universities. When given the choice, university-based engineers, like their counterparts in science, are more apt to pursue technically sweet projects than those that are primarily of economic value, and this preference can powerfully influence the values of those studying in such institutions. But since practically all university research in science and engineering now requires some form of outside sponsorship, research on economically useful projects will receive more attention when the number of technically sweet projects is limited. Such is the case at present, and there is reason to think that the next generation of engineers will be somewhat more attuned to the marketplace than the generation that received their degrees during the decades in which government projects dominated university-based research.

Charles Schaffner made this point most emphatically when he said:

> The engineering curricula of today, the products of the engineering schools, the growth of the faculty and of faculty types, and the directions and everything that was created following World War II, all stem directly from federal government decisions in terms of first, defense, and second, NASA. These programs drenched the engineering schools with research money and pushed them in a direction that had nothing to do, in essence, with the business of the citizenry other than its defense.

Eugene Merchant has concurred with this assessment, saying that "the Apollo program really finished off what the heavy Department of Defense support for research in universities started, namely, turning university engineering research and education away from an orientation towards civilian industry." One consequence of this emphasis, as Aaron Gellman has pointed out, was to decouple the very concept of engineering from normal markets. But as Gellman has also noted, times have changed and now all engineers, including those located in universities, must pay much more attention to the appropriability of their research, for that is what will determine its value in the current market for technological innovation.

Engineering in Society

Engineering is a go-ahead profession, much more given to problem solving than self-reflection. And yet, as the contexts within which engineers operate become more complex and as the interactions between society and engineering become more intricate and constraining, it becomes increasingly important that engineers have a clear understanding of their profession and the ways in which it is connected to the larger society of which it is a part. In an earlier era, when the practice of engineering was largely an autonomous activity, one could afford to defer such reflections until retirement or bash them out on short notice when called upon to address an audience eager to celebrate the achievements of the profession. But today the absence of a carefully documented and fully reasoned justification for positions taken creates a vulnerability that may result in real harm, especially in the competition for good students and research support, and at the very least reflects badly on the profession. This is both unfortunate and unnecessary, for the case for the importance of engineering, when well presented, is quite compelling.

The critical examination and reconceptualization of one's collective identity is a demanding task, one that only those who believe in themselves can successfully complete. But engineers are particularly well situated in this regard, for what other profession is of comparable importance in contemporary society? What is called for then is not a defense of the legitimacy of engineering, and certainly not a public-relations style puffing of its achievements, but rather a patient, evidentially grounded examination of the ways in which engineering functions in contemporary society. The key here is to see engineering as a distinct activity in society, not as an autonomous enterprise that on occasion acknowledges its tenuous connections to society. In recent years the profession of medicine has been subjected to a detailed and sometimes painful demythologizing, one consequence being that today it is widely recognized that medicine is a technical enterprise conducted under strong social constraints and having important social consequences. Engineering is in many ways like medicine, and while it may be able to avoid the more extreme forms of criticism that have been directed at physicians and their organizations, it will in time come to be understood primarily in terms of its functional role in society. Humanists and social scientists who study technology and engineering have already made a beginning in this direction, but to date their efforts have had little impact within engineering itself. In any case, primary responsibility for this effort must remain with the engineers, for it is their self-perception and public image that are at stake.

The dangers of leaving the public interpretation of engineering entirely to others is nicely illustrated by the relationship between the contemporary aesthetic doctrine of postmodernism and engineering, a relationship that Thomas Hughes has reflected on at some length. Postmodernism is a reaction to the twentieth-century cultural style called modernism, a style that since its formulation early in the twentieth century has profoundly influenced all aspects of design from the sculpting of furniture to the planning of cities. The early modernists seized on what they took to be the defining feature of engineering, namely, its efficient use of materials and energy, and declared this to be the fundamental principle of modern aesthetics as well. Modernist architects insisted that less is more, that is to say that beautiful objects are made with a minimum of material and a simplicity of design, and that form follows function. Engineers could not help but find such a doctrine appealing, for it not only honors design values central to engineering, it elevates those values to the level of high art. Indeed, what could be more flattering to engineers than to have designers, and especially architects, treat them not merely as producers of goods but rather as creators of profoundly humane and beautiful objects. They thus had little reason to criticize the public identification of modernism and engineering, even though if pressed most engineers would have admitted that the doctrines of modernism focus on only one aspect of their profession.

Postmodernists, as Hughes points out, stand in complete opposition to what they consider the sterility of modernism. Unwilling to accept what they see as the diminishing constraints of the modernist movement, the postmodernists. reject the primacy of material efficiency in favor of a more varied and accommodating aesthetic. Robert Venturi, the earliest and most articulate of the postmodernists, asserts that "less is not more, less is a bore." He rejects the image of the architect-engineer as a heroic builder and dismisses Le Corbusier's proposal for leveling Paris to clear the ground for a new Cartesian city by saying that architecture "must embody the difficult unity of inclusion rather than the easy unity of exclusion." Instead of geometric fortresses unencumbered by suburbs, Venturi favors "messy vitality."

Why should engineers be concerned with this debate? At the very least they should be aware that many people outside their profession, and especially those concerned with questions of design, creativity and art, see the modernist/ postmodernist debate as, among other things, an examination of the place of engineering in modern society. In this debate the modernists have been allowed to define what engineering is and, as we have seen, their definition is at best a partial one. It ignores the vital linkage between engineering values and market values that has been characteristic of engineering practice throughout this century. Had this linkage been recognized, the "postmodernist" automobiles created by Sloan's designers to realize the strategy of the annual model change would be seen to be just as much a product of modern engineering as was Ford's Model T. As things now stand, however, the postmodernists see no reason not to accept the modernist's identification of their doctrines with the essence of engineering, and engineers feel they have been treated unfairly when told they don't know how to deal with messy vitality. If they wish to prevent such misrepresentations and misunderstandings in the future, engineers ought to be more attentive to the ways in which their profession is presented to the public at large.

What it means to be a professional engineer also needs to be reconceptualized. Living as we do in the age of mass professionalism, in which nearly every occupation has been transformed, at least in name, into a profession, simply asserting that one is a professional is not very informative. Being a professional no longer entails sharing a common culture, since today cultural preferences and practices are largely matters of personal choice. Nor does it signify, in any discriminating sense, being educated, for today nearly half those of college age are enrolled in degree programs of one sort or another. Had professional societies been more vigorous in exercising self-discipline, the concept of professional behavior might be more meaningful than it is today, but such has not been the case. And had colleges and universities been as concerned with the economic health of the professions as they have been with their own expansion, we might be able to say that a professional is someone who enjoys the advantages associated with limited access to privileged status. The compromising of these older meanings of the concept of profession does not, of course, render meaningless the engineer's striving for professionalism. But the nature of the goal sought needs to be redefined in ways that are informative both to engineers and to those who worry about how the profession of engineering serves society at large.

The ultimate goal of all such reconceptualizations is to develop within the community of engineers an increased ability to perceive, describe, and manage the diversity of modern engineering and the ways it changes in time. Engineering is a dynamic enterprise, both internally and in its relations with other aspects of society. As new specialties emerge, new attitudes toward work and management appear, new techniques of design and production are developed, and new expectations gain in importance, engineers need to be able to understand the forces that bring about these changes and the ways in which they can be integrated into existing patterns of thought and behavior. By knowing themselves better, engineers will be better able to serve their profession and its larger purposes successfully.

CONCLUSIONS AND RECOMMENDATIONS

The Resilience of the Engineering Manpower System

Conclusions

1. Examination of previous crises in the engineering manpower system suggests that it has responded adequately and that calls for a radical expansion or reconstruction of existing arrangements for educating engineers cannot be justified by appeals to past experience.
2. Engineers have in the aggregate adapted rapidly and successfully to sudden changes in the demand for particular engineering specialties. Their ability to do so is directly dependent upon their mastery of the fundamentals of design and their knowledge of the underlying mathematics and science.

Recommendations

1. The technical/scientific content of the undergraduate engineering curriculum should emphasize science, mathematics, and engineering design. Technical courses focusing on problems associated with particular engineering specialties should occupy a secondary position in all engineering curricula.
2. When introducing new technologies that render obsolete the knowledge and skills of engineers already employed, companies have an obligation to provide these engineers with educational opportunities that will enable them to remain productive. The continuing education programs offered by many colleges and universities may be helpful in this regard.

The Conceptualization and Presentation of Engineering

Conclusions

1. The ways in which engineering is presented to and understood by the general public is a matter of vital concern to engineers.
2. The nature of engineering can only be understood in a comprehensive manner if its many links to other sectors of society are described and analyzed in a detailed and careful way.

Recommendations

1. The social/humanistic component of the engineering curriculum should concentrate on issues and subjects of direct concern to engineers and interpret them by using the insights and analytic techniques of the social sciences and humanities. Courses such as the History of Technology, Ethics for Engineers, and Engineering and Public Policy offer valuable means for ensuring that engineering students will gain some understanding of the complex contexts of contemporary engineering.

2 Engineers, with the help of historians, philosophers, and other humanists and social scientists, should organize and encourage scholarly studies and public presentations designed to explicate the nature of engineering in all its many different forms. Studies of the interactions between engineering and other sectors of modern society and culture should be especially encouraged.

ACKNOWLEDGMENTS

This report is an attempt to weave together and draw appropriate lessons from the historical papers and comments presented at a three-day conference sponsored by the National Research Council. The author is grateful to all those who prepared the thematic and case studies that occasioned lively discussion at the conference and provide the substantive content of this report. While the report draws heavily upon the proceedings of the conference, it does not attempt to provide an exact summary of what took place at that meeting, and the conference participants are in no way responsible for either the general conclusions and recommendations of this report or such errors as have been introduced during its preparation. The author is also extremely grateful to those individuals who took the trouble to review an earlier draft of this report, and especially to Melvin Kranzberg, Samuel Florman, Courtland Lewis, and Edwin Layton. Their constructive suggestions have improved the report at many points, while such errors and misjudgments as it still contains remain solely the responsibility of the author. Finally, the author would like to thank the staff and members of the Committee on the Education and Utilization of the Engineer, and especially George Ansell, for giving him the opportunity to participate in and contribute to their activities. It has been a valuable and much appreciated experience, and a most unusual one for an historian.

PARTICIPANTS, CONFERENCE ON ENGINEERING INTERACTIONS WITH SOCIETY

Presenters

James E. Brittain, School of Social Sciences, Georgia Institute of Technology, "Engineering in Industrial Research and Development"

P. Thomas Carroll, Division of Science and Technology Studies, Rensselaer Polytechnic Institute, "Orphaned Innovations: The Development of Large-Scale Solid Rocket Boosters at the Jet Propulsion Laboratory"

Edward W. Constant II, Department of History, Carnegie Mellon University, "Technological Knowledge about Engineering Manpower: Some Preliminary Considerations"

Eugene Ferguson, University of Delaware and Hagley Museum, Panelist

Samuel Florman, Kreisler, Borg, Florman Construction Company, New York, Panelist

Robert Friedel, IEEE Center for the History of Electrical Engineering, "Engineers and the Micro Revolution: The Emergence and Impact of Solid-State Electronics"

James Hansen, Historian for NASA, Langley Research Center, "The Revolt against Max Munk at Langley Aeronautical Laboratory: A Case Study of the Fate of an Eccentric in an American Engineering Community"

David A. Hounshell, Curator of Technology, Hagley Museum, and Department of History, University of Delaware, "Redesigning Production Engineering: Mass Production and the Model Change"

Thomas P. Hughes, Department of History and Sociology of Science, Technology and Medicine, University of Pennsylvania, Panelist

Melvin Kranzberg, Callaway Professor of the History of Technology, Georgia Institute of Technology, "Engineering Education and Sociotechnical Needs: Reaction and Interaction"

Larry Lankton, Department of Science, Technology and Society, Michigan Technological University, "The Social Side of Early American Engineering"

Stuart W. Leslie, Mellon Scholar in the History of Science, Johns Hopkins University, "Industrial Research and Product Development at General Motors"

Michal McMahon, Historical Consultant and Department of Humanities and Communication, Drexel University, "Engineering Education as 'Best Practice': Historical Reflections on the Crisis"

Nathan Reingold, Editor, Joseph Henry Papers, Smithsonian Institution, "Vannevar Bush, Applied Mathematics, and the Nature of Engineering"

Martin Reuss, Civil Works Historian, U.S. Army Corps of Engineers, "Politics, Technology and the Development of Hydraulic Engineering: The Influence of Andrew A. Humphreys"

Alex Roland, Department of History, Duke University, "The Race to the Moon: The Experience at NASA"

Jeffrey L. Sturchio, Department of Humanities, New Jersey Institute of Technology, "Crisis in Industrial Chemistry: Synthetic Organic Chemicals and World War I"

Neil Wasserman, Research Associate, Harvard Business School, "The Development of an Engineering Organization at AT&T"

Committee Members*

George S. Ansell, Dean of Engineering, Rensselaer Polytechnic Institute, and Chairman, Panel on Engineering Interactions With Society

Jordan J. Baruch, President, Baruch Associates, Washington, D.C.

Erich Bloch, Vice-President, Technical Personnel Development, IBM Corporation

Dennis Chamot, Assistant Director, Department of Professional Employees, AFL-CIO

Aaron J. Gellman, President, Gellman Research Associates, Inc., Johnstown, Pa.

Helen Gouldner, Professor of Sociology and Dean, College of Arts and Sciences, University of Delaware

Jerrier A. Haddad, Chairman and Study Director, Committee on the Education and Utilization of the Engineer

Lawrence M. Mead, Jr., Senior Management Consultant, Grumman Aerospace Corporation

M. Eugene Merchant, Principal Scientist, Manufacturing Research, Cincinnati Milacron, Inc.

Robert M. Saunders, Acting Dean, School of Engineering, University of California, Irvine

*Titles and affiliations are as of the time of the conference.

Charles E. Schaffner, Executive Vice-President, Syska and Hennessey, New York, N.Y.

Judith A. Schwan, Assistant Director, Research Laboratories, Eastman Kodak, Inc.

Donald G. Weinert, Executive Director, National Society of Professional Engineers

Other

Arthur L. Donovan, Director, Center for the Study of Science in Society, Virginia Polytechnic Institute and State University, Conference Chairman and Rapporteur

Gary L. Downey, Assistant Professor of Science and Technology Studies, Center for the Study of Science in Society, Virginia Polytechnic Institute and State University

Lewis G. (Pete) Mayfield, Head, Office of Interdisciplinary Research, National Science Foundation

William H. Michael, Jr., Executive Director, Committee on the Education and Utilization of the Engineer

Vernon H. Miles, Assistant Director, Committee on the Education and Utilization of the Engineer

Herbert H. Richardson, Head, Department of Mechanical Engineering, Massachusetts Institute of Technology

Steve Tucker, Program Manager, Edison Engineering Program, General Electric